PHYSICS RESEARCH AND TECHNOLOGY

GRAVITOELECTROMAGNETIC THEORIES AND THEIR APPLICATIONS TO ADVANCED SCIENCE AND TECHNOLOGY

PHYSICS RESEARCH AND TECHNOLOGY

PHYSICS RESEARCH AND TECHNOLOGY

GRAVITOELECTROMAGNETIC THEORIES AND THEIR APPLICATIONS TO ADVANCED SCIENCE AND TECHNOLOGY

T. MUSHA
M. J. PINHEIRO
AND
T. VALONE

NOVINKA
New York

For permission to use material from this book please contact us:
Telephone 631-231-7269; Fax 631-231-8175
Web Site: http://www.novapublishers.com

Library of Congress Cataloging-in-Publication Data

ISBN: 978-1-62808-210-4

Published by Nova Science Publishers, Inc. † New York

CONTENTS

PREFACE

Gravitoelectromagnetism refers to a set of formal analogies between the equations for electromagnetism and relativistic gravitation;, specifically between Maxwell's field equations and an approximation, valid under certain conditions, to the Einstein field equations for general relativity. Gravitomagnetism is a widely used term referring specifically to the kinetic effects of gravity, in analogy to the magnetic effects of moving electric charge. The analogy and equations differing only by some small factors were first published in 1893 by Oliver Heaviside as a separate theory expanding Newton's law. T.T.Brown, who firstly discovered the electrogravitic effect, performed several experiments during 1950s and succeeded to generate thrust without the reliance on a surrounding medium by applying high voltages to materials with high dielectric constants. In Tomas Valone's book, "Electrogravitic Systems", published in 1994, P.A. LaViolette claimed that electrogravitic technology was developed under U.S. Air Force black project since late 1954, and has alledged that is has been put to practical use in the B-2 Advanced Technology Bomber to provide an exotic auxiliary mode of propulsion.

This book outlines the physical basis of gravitoelectromagnetic theories and related physics which might lead to the development of advanced technologies for future aviation systems and other purposes. The intent of the present volume is to assist the reader in the comprehension of the mathematical basis of gravitoelectromagnetism and its application..

Chapter 1 introduces the historical background of electrogravitics, especially the Biefeld-Brown effect. The second chapter provides explanations of the Biefeld-Brown effect and other related phenomena. The third chapter is concerned with the electrogravitic effect as related to zero point energy fluctuation in the vacuum from the standpoint of quantum electrodynamics.

The fourth chapter is concerned with other electromagnetic gravity control devices including Heim's theory and the Li-Torr electrogravitic theory. The fifth chapter is an exposition that the Abraham force is the analogue of the Magnus force, and it thus represents the formation of a vortex structure with an electromagnetic nature in the physical vacuum: it is therefore analogous to an electromagnet toroid that can generate gravitational field. The sixth chapter deals with a plasma theory of the Universe and the role played by the gravito-electromagnetic forces generated by the plasma permeating the space between planets. And the last chapter shows the application on advanced aviation systems and future prospect of these technologies.

This is a textbook written for both students and professional scientists, providing the mathematical basis for readers to introduce the basic concept of gravitoelectromagnetic theories, and also providing elementary applications.

Our sincere hope is that this book will provide sound guidance for researchers and students to further explore research in the realm of field propulsion theory and possible applications.

ACKNOWLEDGMENTS

The authors appreciate Paul Murad (ret. US Department of Defense), Gary V. Stephenson (Seculine Consulting) for their help and suggestions, which greatly improved this book.

Chapter 1

HISTORY OF ELECTROGRAVITICS

Thomas F. Valone

Integrity Research Institute, Beltsville, MD, US

ABSTRACT

The Biefeld–Brown effect was initially investigated by Thomas Townsend Brown and Dr. Paul Alfred Biefeld in the 1920s. Research continued through the 1950s and 1960s by Brown and other researchers. The use of this electrogravitic propulsion effect was further explored during the publicized era of gravity control propulsion research, which included the United States gravity control propulsion initiative. The effect was almost certainly included in the research conducted at Martin Aircraft's RIAS (Research Institute for Advanced Studies) unit in Baltimore during this period, which is reported to have been specifically interested in gravity, and which hired Burkhard Heim as a consultant because his "Heim Theory" appeared to have significant implications for the nature of gravity.

1. INTRODUCTION

"Electrogravitics" is the science of using high voltage electricity to provide propulsive force to aircraft or spacecraft of certain geometries. Or as Jeane Manning explains, "The apparatus is pulled along by its self-generated gravity field, like a surfer riding a wave." Its discovery is often credited to

Thomas Townsend Brown, a physicist who was encouraged by his professor, Dr. Paul Biefeld, a former classmate of Albert Einstein. However, there are those who say that Professor Francis Nipher's experiments, electrically charging lead balls, published in the *Electrical Experimenter* in 1918, are the predecessor to Biefeld/Brown, offering an insight to the inception of the phenomenon.

Recently, two 1956 military documents, "Electrogravitics Systems" and "The Gravitics Situation," originally published by the Gravity Research Group of London (Special Weapons Study Unit), were declassified from a “confidential” status. Outlining T. Townsend Brown's antigravity discovery [1], and the subsequent Project Winterhaven, they were a vital new chapter in aviation research. For example, the documents state, "Unlike the turbine engine, electro-gravitics is not just a new propulsion system, it is a new mode of thought in aviation and communications, and it is something that may become all-embracing"[2]. Further advances have been made since then, warranting a second volume to the series as noted elsewhere in this chapter.

2. Researchers and Experiments on Electrogravitics

Francis Nipher

Dr. Francis Nipher, Professor of physics, Washington University, St. Louis, Missouri, did some of the pioneering electrogravitics work at Washington University in St. Louis back around the turn of the last century. He applied high voltage to lead balls, lead spheres and hollow metal boxes and compared the repulsive effect induced in small test spheres hung vertically near them, similar to the original Cavendish experiments but with high voltage. Dr. Nipher went to great lengths to insert protective, grounded screens of glass between the solid lead spheres and the suspended balls to rule out electrostatic effects.

Before connecting any form of electric current to the modified Cavendish apparatus, Prof. Nipher took special precaution to carefully screen the moving element from any electrostatic or electromagnetic effects. His apparatus briefly consists of two large lead spheres ten inches in diameter, resting upon heavy sheets of hard rubber. Two small lead balls, each one inch in diameter, were now suspended from two silk threads, stationed at the sides of the two large lead spheres, from which they were separated by a little distance. Moreover, the suspended balls were insulated elaborately from the large spheres by enclosing them first airtight in a long wooden box, which was also covered with tinned iron sheets as well as cardboard sheets. There was, furthermore, a metal shield between the box and the large metal spheres. The large metal lead spheres now exerted a certain gravitational pull upon the suspended small lead balls ... and the small lead balls were slightly pulled over towards the large spheres. In further experiments Prof. Nipher decided to check his results. To do this he replaced the large solid lead spheres with two metal boxes, each filled with loose cotton batting. These hollow boxes (having practically no mass) rested upon insulators. They were separated from the protective screen by sheets of glass and were grounded to it by heavy copper wires. The metal boxes were then charged in every way that the solid lead spheres had been, but not the slightest change in the position of the lead balls could be detected. This would seem to prove conclusively that the "repulsion" and "gravitational nullification" effects that he had produced when the solid balls were electrically charged were genuine and based undoubtedly on a true inter-atomic electrical reaction, and not upon any form of electrostatic or electromagnetic effects between the large and small masses. If they had been, the metal boxes, with no mass, would have served as well as the solid balls. (excerpts from [3])

Professor Nipher had previously published his theory on "Gravitation and Electrical Action" with extensive equations in his papers[4,5], building on the suggestions of an "intimate relation between gravitation and electrical action in several other journals[6-8]. Most of the cited articles are available on the internet (credit: RexResearch.com) and are a fascinating inquiry into the early anomalies with high voltage effects on propulsion, as well as the difference between the negative polarity versus positive polarity electrogravitics.

T. Townsend Brown

> "The long and constant persuasion that all the forces of nature are mutually dependent, having one common origin, or rather being different manifestations of one fundamental power, has often made me think on the possibility of establishing, by experiment, a connection between gravity and electricity ...no terms could exaggerate the value of the relation they would establish." -- Michael Faraday, physicist, 1865.

Of all the names of 20th Century science, perhaps none is more obscure, or more curious, than that of Thomas Townsend Brown.

Brown was blessed with the unique ability to "see what others have seen and think what nobody has thought." As a teenager in the 1920s, working in a well-equipped laboratory in the basement of his prominent Ohio family's opulent home, Brown noticed an unusual effect when high voltage was applied to a Coolidge X-Ray tube. With that observation, he came to believe he had discovered a link between electricity and gravity - and a way to lift and propel flying vehicles by purely electrical means.

Thus begins the odyssey of T. Townsend Brown, who spent a lifetime crisscrossing the hemisphere in the relentless pursuit of his "flying saucer pipe dreams." The first publication of the research Brown conducted was revealed in the now famous article, "How I Control Gravity," *Science and Invention*, Aug 1929. It is reprinted in its entirety with detailed diagrams of each experiment in *Electrogravitics II* [9]. However, it wasn't until the publication of this second anthology on electrogravitics that this author questioned the "problem" that Brown reported in the 1929 article. Brown indicated that his high voltage impulse exhausts itself rapidly and the mass being charged returns to its original position invariably, without concluding what seems obvious today that the impulse or *changing current* is the prime mover. Thanks to definitive work by Professor Oleg Jefimenko, such an "electrokinetic" effect could be discerned as being different from the original concept of electrogravitics, all of which is explained in detail in the 2000 publication of the second volume of electrogravitics.

Brown and his colleagues anticipated the situation even as early as 1923, and began at that time to construct the necessary theoretical bridge between the two then separate phenomena, electricity and gravitation. The first actual demonstration of the relation was made in 1924.

Dr. Mason Rose

An interesting summary of the early work of T. Townsend Brown in found in the article entitled, "The Flying Saucer: The Application of the Biefeld-Brown Effect to the Solution of the Problems of Space Navigation" by Mason Rose, Ph.D., President University for Social Research (1952). It was published in two magazines, *Science and Invention*, August 1929, and *Psychic Observer*, Vol. XXXVII, No. 1, 1929. Below is an excerpt from the beginning of the article:

The scientist and layman alike encounter a primary difficulty in understanding the Biefeld-Brown effect and it's relation to the solution of the flying saucer mystery. A proper interpretation of this theory is prevented because both scientist and layman are conditioned to think in electromagnetic concepts, whereas the Biefeld-Brown effect relates to electrogravitation. Their lack of awareness is justifiable, however, because the data on electrogravitation, inasmuch as it is a comparatively recent and unpublished development, has limited availability and circulation.

Townsend Brown, the discoverer of electrogravitational coupling, is the only known experimental scientist in this new area of scientific development as of this writing. Thus, anyone wanting to understand electrogravitation and its applications to astronautics must dismiss the principles of electromagnetics in order to grasp the essentially different principles of electrogravitation. Electrogravitational effects do not obey the known principles of electromagnetism. Electrogravitation must be understood as an entirely new field of scientific investigation and technical development.

The most efficient method of effecting an understanding of electrogravitation is to review the evolutionary development of electromagnetism. From the smallest atom to the largest galaxy, the universe operates on three basic forces, namely Electricity, Magnetism and Gravitation. These forces can be represented shown as follows:

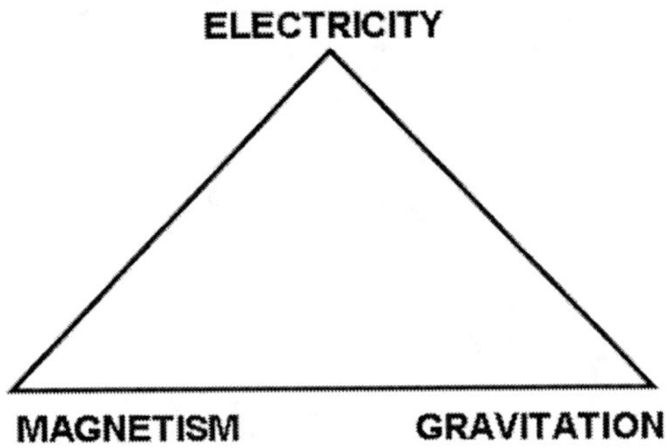

Figure 1. Fundamental Three Forces.

Taken separately, these forces are of no real practical use. Electricity by itself is static electricity and therefore functionless. It will make your hair stand on end, but that is about all. Magnetism by itself has very few practical applications aside from the magnetic compass, and gravity simply keeps objects and people pinned to the earth. However, when they are used to work in combination with each other, almost endless technical applications come into being. Currently, our total electrical development is based on the coupling of electricity with magnetism, which provides the basis for the countless uses we make of electricity in modern societies. Faraday conducted the first productive empirical experiment with electromagnetism around 1830, and Maxwell did the basic theoretical work in 1865.

The application of electromagnetism to microscopic and sub-microscopic particles was accomplished by Max Planck's work in quantum physics about 1890; and then in 1905 Einstein came forward with relativity, which dealt with gravitation as applied to celestial bodies and universal mechanics. It is principally out of the work of these four great scientists that our electrical developments, ranging from the simple lightbulb to the complexities of nuclear physics, have emerged. In 1923, Dr. Biefeld, Professor of Physics and Astronomy at Dennison University and a former classmate of Einstein in Switzerland, suggested to his protoge, Townsend Brown, certain experiments which led to the discovery of the Biefeld-Brown effect, and ultimately to the electrogravitational energy spectrum (in actuality, it was Brown who first observed the effect and brought it to the attention of Dr. Biefeld, who suggest further experiments to determine the origin of and enhance the effect - Juniper). Biefeld wondered if an electrical condenser, hung by a thread, would have a tendency to move when it was given a heavy electrical charge. Townsend Brown provided the answer. There is such a tendency.

After 28 years of investigation by Brown into the coupling effect between electricity and gravitation, it was found that for each electromagnetic phenomenon there exists an electrogravitational analogue. This means, from the technical and commercial viewpoint, potentialities for future development and exploitation are as great or greater than the present electrical industry. When one considers that electromagnetism is basic to the telephone, telegraph, radio, television, radar, electric generators and motors, power production and distribution, and is an indispensable adjunct to transportation of all kinds, one can see that the possibility of a parallel, but different development in electrogravitation has almost unlimited prospects.

The initial experiments conducted by Townsend Brown, concerning the behavior of a condenser when charged with electricity, had the characteristic of simplicity which has marked most other great scientific advancements. The first startling revelation was that if placed in free suspension with the poles horizontal, the condenser, when charged, exhibited a forward thrust toward the positive poles. A reversal of polarity caused a reversal of the direction of thrust. The experiment was set up as follows:

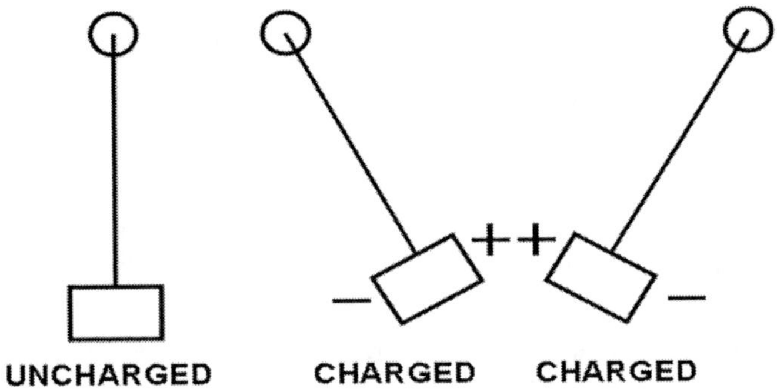

Figure 2. Biefeld-Brown Effect experimental set-up.

The antigravity effect of vertical thrust is demonstrated by balancing a condenser on a beam balance and then charging it. After charging, if the positive pole is pointed upward, the condenser moves up. If the charge is reversed and the positive pole pointed downward, the condenser thrusts down.

These simple experiments demonstrate what is now known as the Biefeld-Brown effect. It is the first and, to the best of our knowledge, the only method of affecting a gravitational field by electrical means. It contains the seeds of control of gravity by man. The intensity of the effects is determined by five factors, which are:

1. The separation of the plates of the condenser, the closer the plates, the greater the effect.
2. The ability of the material between the plates to store the electrical energy in the form of elastic stress. A measure of this ability is called the "K" of the material. The higher the "K," the greater the Biefeld-Brown effect.
3. The area of the plates, the greater the area giving the greater effect.
4. The voltage difference between the plates; more voltage, more effect.
5. The mass of the material between the plates; the greater the mass, the greater the effect.

It is this fifth point which is inexplicable from the electromagnetic viewpoint and which provides the connection with gravitation. On the basis of further experimental work from 1923 to 1926; Townsend Brown in 1926, described what he called a "space car." This was a revolutionary method of terrestrial and extraterrestrial flight, presented for experiment while motor propelled planes were yet in a primitive stage. This engineering feat by Townsend Brown was all the more remarkable when we consider such a machine produces thrust with no moving parts, does not use any aerodynamic principles of flight, and has neither control surfaces, or a propeller. Townsend Brown had discovered the secret of how the flying saucers fly years before and such objects were reported.

Now the basic differences between electromagnetism and electrogravity have been described and the basic principles of the Biefeld-Brown effect have been outlined, we are finally ready to understand the principles of astronautics or the conquest of space. The earth creates and is surrounded with a gravitational field which approaches zero as we go far into space. This field presses objects and people to the earth's surface; hence it presses a saucer object to the earth.

However, through the utilization of the Biefeld-Brown effect, the flying saucer can generate an electrogravitational field of its own which modifies the earth's field. This field acts like a wave, with the negative pole at the top of the wave and the positive pole at the bottom, the saucer travels like a surfboard on the incline of a wave that is kept continuously moving by the saucer's electrogravitational generator.

Since the orientation of the field can be controlled, the saucer can thus travel on its own continuously generated wave in any desired angle or direction of flight. Since the saucer always moves towards its positive pole, the control of the saucer is accomplished by varying the orientation of the positive charge. Control, therefore, is gained by switching charges rather than by the control surfaces. Since the saucer is traveling on the incline of a continually moving wave which it generates to modify the earth's gravitational field, no mechanical propulsion is necessary.

T. Townsend Brown's Experiments

Only a brief summary of the experiments of T. Townsend Brown are possible here since they are extensive and only a chapter is available for the entire history.

During the early years of the 20th century, starting in 1927, Brown patented a series of designs of various craft. The only time that "gravitators", gravitation, and electrogravitics is mentioned however, is in the first patent

Brown obtained in England (#300,311). All of the rest for the next three decades referred to "electrokinetics" instead. It is still a matter of speculation as to the difference between the two phenomena. However, in the article "How I Control Gravity", Brown refers to many experiments where the impulse of current is abrupt and the force is only momentary (article is reprinted in its entirety in *Electrogravitics II* by T. Valone). As mentioned earlier, this is a key to the first derivative effect of the change in current over time discovered by Dr. Oleg Jefimenko, who also called it the "electrokinetic force" in his book explaining the same phenomenon [10]. Brown secured several more patents in the period 1930-1970 and all of them refer to "electrokinetics" rather than electrogravitics. The culmination of the solo Brown experiments was the presentation on March 27, 1952 for the Office of Naval Research. An entire report was generated for this demonstration entitled, "The Electro-Gravity Device of Townsend Brown", File: 24-185, Subject: Special Inquiry, (available in its entirety from Integrity Research Institute). Perhaps the most entertaining part of the report is the transcript of a telephone conversation between a Major General and a Lt. General about the event. The Major General was "frightened" since the entire saucer demonstration was being conducted by a "private group".

From 1958 to about 1961, Agnew Bahnson formed a partnership with T. Townsend Brown and conducted extensive experiments on larger saucer-shaped craft. Integrity Research Institute obtained the rights to a silent video now on DVD featuring segments from the Brown-Bahnson laboratory experiments with calendar dates showing in the scenes. Several scenes show hovering craft under the influence of high voltage. A four minute segment is also available on the internet under the title, "T.T. Brown Electrogravity Vacuum Experiments – YouTube". More information about these early experiments is available, including all of the Brown and Bahnson patents, from *Electrogravitics Systems* and from *Electrogravitics II*, by Thomas Valone.

T. T. Brown's 1955-1956 Paris Experiments Revealed

In 1955 and 1956 Townsend Brown made two trips to Paris where he conducted tests of his electrokinetic apparatus and electrogravitic vacuum chamber tests in collaboration with the French aeronautical company Société National de Construction Aeronautiques du Sud Ouest (S.N.C.A.S.O.) . He was invited there by Jacques Cornillon, the company's U.S. technical representative. The project was named Project Montgolfier in honor of the two French brother inventors who performed early aircraft flights. The project continued for several years until the company changed ownership resulting in a final report which was written up in 1959 [11].

Details of the Project Montgolfier experiments remained a closely guarded secret for many years until Jacques Cornillon courageously decided to make them public prior to his death in July 2008. Brown's proposal, the project's top secret final report, and an assortment of revealing diagrams and photos are posted on the Cornillon website at: Project Montgolfier: http://projet montgolfier.info/and are available for free download. Brown's proposal is in English, whereas the secret Montgolfier Project final report is in French. An English translation of this 100 page final report is available from the Peeteelab website. (Note there is a translation error in the statement of the electrokinetic disc wire size, which is 100 times smaller than quoted. Refer to the original French document for the proper size.) The flying disc carousel experiment that the Montgolfier Project conducted in 1955 used 2-1/2 foot diameter discs (75 cm dia.) hung from 4 meter tethers suspended from the ends of a 3 meter arm. Based on the description given, this seems to have been almost the same flying disc test that Brown gave to the Navy at Pearl Harbor a year or two earlier.

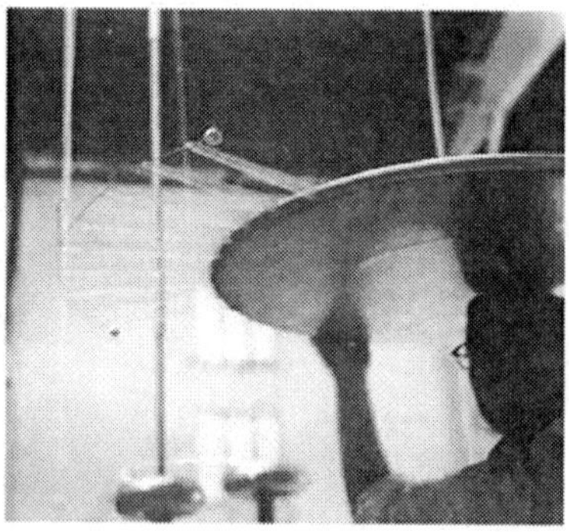

Figure 3. Left: Brown holding a flying disc tested in Project Montgolfier. Right: Close-up of disc showing outboard leading-edge wire. (photos courtesy of J. Cornillon).

Based on the angle of the disc suspension cable seen in the photo on the right below, one may estimate that the disc was traveling at a speed of ~8.7 meters per second, or about 20 mph. It would have completed one revolution of its 18 meter course in 2 seconds.

Figure 4. Left: Carousel test rig. Right: Disc in flight. (photos courtesy of J. Cornillon).

Brown had finished his collaboration with S.N.C.A.S.O. in 1956. From a letter that Mr. Cornillon later wrote to a colleague, we learn that in October 1957 Brown was in the process of test flying 10 foot diameter discs energized at a voltage of 300 kV! Here we see that Brown had followed the plan he had first set out in his 1952 Project Winterhaven proposal which was to eventually test fly a ten foot diameter disc powered by 500 kV (70% more voltage than he used in his 1957 test flight). Hence we see that by this early date Brown had progressed beyond the toy model stage to flying small scale aircraft. To reach this stage he must have been receiving substantial funding from either the military or from a major corporation. More about Project Winterhaven and Brown's research may be found in the book *Secrets of Antigravity Propulsion*, by Paul LaViolette [12].

In addition the Project Montgolfier team constructed a very large vacuum chamber for performing vacuum tests of smaller discs at a pressure of 5×10^{-5} mm Hg; see below.

In reading the section describing the vacuum chamber results, we learn that when the discs are operated at atmospheric pressure they move in the direction of the leading edge wire regardless of outboard wire polarity. This indicates that in normal atmospheric conditions the discs are propelled forward primarily by unbalanced electrostatic forces due to the prevailing nonlinear field configuration (which causes thrust in the direction of the low field intensity ion cloud regardless of the ion polarity). On the other hand, the report says that under high vacuum conditions the discs always moved in the direction of the positive pole, regardless of the polarity on the outboard wire. This indicates that in the absence of the unbalanced forces exerted by ion clouds, the discs moved mainly on the basis of the electrogravitic field effect, always toward the positive (negative G) direction. These vacuum chamber experiments were a decisive milestone in that they demonstrated beyond a doubt that electrogravitic propulsion was a real physical phenomenon. The report concludes saying: "It seems perfectly reasonable to conclude that a concentrated force of some kind accumulates within the presence of a strong dielectric." (i.e., presumably in the presence of a high-K dielectric.)

Figure 5. Left: Vacuum chamber vessel (1.4 m diameter) for conducting electrogravitic tests. Right: Vessel opened to show test rotor rig within. (Photos courtesy of J. Cornillon).

Electrogravitics Systems: An Examination of Electrostatic Motion Dynamic Counterbary and Barycentric Control, Gravity Research Group, 1956

At the same time the France experiments were conducted, a report surfaced as the first of two reports published by the Aviation Studies Ltd. in London, which still is in existence today. It is reprinted in *Electrogravitics*

Systems: Reports on a New Propulsion Methodology and summarizes in simple form the work that has been done and is being done in the new field of electrogravitics. It also outlines the various possible lines of research into the nature and constituent matter of gravity, and how it has changed from Newton to Einstein to the modern Hlavaty concept of gravity as an electromagnetic force that may be controlled like a light wave. A curious fact revealed in T.T. Brown's first article "How I Control Gravity" (Science and Invention, 1929) is the alignment of the "molecular gravitors". These massive dielectrics provided the most propulsive force when the "differently charged elements" were aligned (with the voltage source). This sounds like crystal plane alignment and perhaps explains the article "Gravity Nullified: Quartz Crystals Charged by High Frequency Currents Lose Their Weight" which appeared two years earlier in the same magazine in September of 1927. The editors had a change of heart however, in the following issue, and rescinded the article.

T. T. Brown's first patent, #1,974,483 issued in 1934, "Electrostatic Motor," is a fascinating free energy machine as well as a propulsion source. Claiming an efficiency of a "million to one", Brown causes the massive dielectrics to be the workhorse of the motor, exceeding, in his words, "the well known pin wheel effect or reaction from a high voltage point discharge." Much of what we know about T.T. Brown is from his numerous patents (all of them are reprinted in the Electrogravitics Systems book), although I was fortunate enough to correspond with him in 1981 when he was at the University of Florida. A sample of his detailed correspondence is contained in the book, Ether-Technology: A Rational Approach to Gravity-Control by Rho Sigma (1977) which is the only other introduction to Brown's work. The important fact from that book is that the DC power supply went up to 250 kV, with a substantial force being displayed starting around 150 kV. Here we get an idea of the range of voltage necessary for successful electrogravitics that even recent military contractors mysteriously disregard. An example is R. L. Talley's report to the Air Force concerned "with exploring the Biefeld-Brown effect which allegedly converts electrostatic energy directly into a propulsive force in a vacuum environment." It was entitled, "Twenty First Century Propulsion Concept" #PL-TR-91-3009, but only tested Brown saucer designs in the range of 19 kV and predictably failed to produce results.

Brown's saucer tests show a propulsive force with the positive voltage leading and the negative edge trailing. The high voltage electrically charged the air around the craft with a cloud of positive ions forming in front of the craft and a cloud of negative ions behind. This has been verified with tests recently performed by researcher Larry Davenport. These tests are reprinted in the book, Electrogravitics Systems and can be seen in the commercial video, "Free Energy, The Race to Zero Point" for which I was the technical consultant.

In March, 1952, the Townsend Brown Foundation laboratory in Hollywood, California was visited by Air Force Major General Victor E. Bertrandias. He reported to Lt. Gen. H. A. Craig that he was "frightened" by the flying demonstration because it was in private hands and he felt it was "in the stage in which the atomic development was in the early days." He expressed concern about "if it ever gets away," meaning, we presume, "into enemy hands." A confidential security investigation was thereby initiated concerning the Foundation and T.T. Brown. Shortly afterwards, an evaluation by the Office of Naval Research in September, 1952 surprisingly devalued the Brown saucers to a "well-known phenomenon of the electric wind" claiming it would perform "less at high voltage and zero in a vacuum." The Navy declassified the report in October, 1952. However, today, copies of the report are not available from the Naval Research Laboratory in Washington, DC.

In 1956, Brown returned from a research trip in France where he verified that electrogravitics worked well in a vacuum, in other words, the environment of outer space. Interavia Magazine published an article in the same year about Brown entitled, "Towards flight without stress or strain or weight"[13] and alluded to possible speeds of several hundred miles per hour. In 1958, Fate magazine writer Gaston Burridge described Brown's metal discs that reached up to 30 inches in diameter. Until 1960, Brown and Agnew Bahnson worked on various designs in Bahnson's laboratory which were recorded by Bahnson's daughter on Super-8 film. Today, a VHS converted silent video is available of those experiments entitled, "Thomas Townsend Brown: Bahnson Lab 1958-1960." In 1964 Bahnson, an experienced pilot, mysteriously flew into electric wires and died. The Bahnson heirs subsequently dissolved the laboratory project.

3. Electrogravitics Puzzle

In 1985, Dr. Paul LaViolette was in the Library of Congress in Washington, DC and looked up the work "gravity" in the card catalog. Surprisingly, he found the listing for "Electrogravitics Systems," a report that was missing from the stacks. When the librarian tried to locate any other copies through interlibrary loan, she commented, "It must be an exotic document" because she could find only one in the country which was at

Wright-Patterson Air Force Base. Thus, LaViolette was successful in obtaining a copy of the formerly classified document. The mystery continued: seven years later when contacting the Wright-Patterson AFB Technical Library, they surprisingly found no reference in the computer-based card catalog. They did locate the document on the shelves, however, after being asked to search for it. To summarize, the report has historic value because:

- It validates T.T. Brown's experiments;
- It lists the major corporations that were collaborating on electrogravitics;
- It includes the requirements for supersonic speed;
- It shows the continuity from Project Winterhaven in 1952;
- The report includes a list of electrostatic patents;
- It had been classified by the Air Force for an undetermined amount of time which underscores its importance.

Prepared by the Aviation Studies (International) Ltd., Gravity Research Group, Special Weapons Study Unit in England in February of 1956, it defines electrogravitics as "a synthesis of electrostatic energy use for propulsion." The report historically notes that: "Electrogravitics had its birth after the War, when Townsend Brown sought to improve on the various proposals that then existed for electrostatic motors sufficiently to produce some visible manifestation of sustained motion." As mentioned in the first section of the report, both Project Winterhaven (1952) and "Electrogravitics Systems" (1956) propose "a saucer as the basis of a possible interceptor with Mach 3 capability." Another interesting detail presented is the necessity of an insulator with a exceedingly high "dielectric" constant of 30,000 for supersonic speed when the best dielectrics of that era were around 5,000. This section goes on to describe the creation of a local gravitational system by the craft which "would confer upon the fighter the sharp-edged changes of direction typical of motion in space." The January, 1955 entry states:

Back in 1948 and 49, the public in the U.S. had a surprisingly clear idea what a flying saucer should, or could do. There has never at any time been any realistic explanation of what propulsion agency could make it do those things, but its ability to move within its own gravitation field was presupposed from its manoeuverability. Yet all this was at least two years before electro-static energy was shown to produce propulsion. It is curious that the public were so far ahead of the empiricists on this occasion..."

The intriguing part of this commentary is that without any space program at the time, the report complains that the public knows how UFOs behave and refers to sharp-edged changes of direction. Later in the report, we read, "One of the difficulties in 1954 and 1955 was to get aviation to take electrogravitics seriously." However, corporations such as Douglas, Sperry, Bell, GE, Hiller, Lear, and Convair are then described with an ongoing-project perspective. For example we read that, "General Electric is working on the use of electronic rigs to make adjustments to gravity." "Glenn Martin say gravity control could be achieved in six years....Clarke Electronics state they have a rig, and add that in their view the source of gravity's force will be understood sooner than some people think." This information makes the report exciting reading and gives it an air of suspense.

Even today, electrogravitics continues to attract public attention in the press. The latest is an article entitled "Military Power" published in a British aviation magazine, *AIR International*, [14] that includes copies of LaViolette's drawings from the Electrogravitics Systems book. The article also cites the Aviation Week and Space Technology article from March, 1992 "Black world engineers, scientist, encourage using highly classified technology for civil applications" which caused LaViolette to investigate the B-2 Bomber connection to T.T. Brown's electrogravitics. Through the investigative effort of Dr. Paul LaViolette, it has become clear that electrogravitics became an integral part of the B-2 Stealth Bomber today, giving it an unlimited range. LaViolette challenges us with the question, "Could the B-2 really be the realization of one of mankind's greatest dreams – an aircraft that has mastered the ability to control gravity?" LaViolette's investigation is summarized in an article "The U. S. Antigravity Squadron" which has been reprinted in his latest book, Secrets of Antigravity Propulsion (Inner Traditions, 2008). Both reports mentioned above, in the book appear in *Electrogravitics Systems, A New Propulsion Methodology*. It is worthwhile mentioning that LaViolette's book, *Subquantum Kinetics: The Alchemy of Creation* includes a chapter on the theory of electrogravitics and a plot of applied voltage versus disc speed from Naval Research Lab data, which starts around 40 kilovolts and 2 miles per hour.

John Searl's Electrogravity

John R.R. Searl, of England, constructed numerous craft purported to fly with high voltage (see the recent biography *Antigravity: The Dream Made*

Reality [15]). However, one correction to the "Antigravity" article from *Atlantis Rising* Number 22 issue is that the positive pole was traditionally at the periphery of Searl's crafts. This is important because as Searl describes his control of the imbalance of positive voltage on the edges, to steer the craft, he found that the saucers would travel toward the more positive side, exactly like T.T. Brown's saucers behave. Throughout the sixties and the seventies, J.R.R. Searl produced many newsletters detailing the work he was doing. Since I corresponded with him in 1981, I also received some of these reports. The importance of his experiments lies in the electrogravitics phenomena associated with them. In the 6/1/68 issue of the "Searl National Space Research Consortium" newsletter, Barrett reports that the ionization of the air and permanent electric polarity of dielectrics were common along with the antigravity effects. In the 6/14/71 issue of the newsletter, Bernhard Vaegs reports that "a pink halo surrounded the craft" and describes the effect of the millions of volts that were generated. This type of description is found throughout the reports and probably was measured by the length of the spark discharge considering the approximate voltage breakdown of air. Barrett describes in the 6/1/68 issue a vacuum layer that surrounds the craft preventing ionizing breakdown of the air. The similarities between Searl's high voltage propulsion and T.T. Brown's high voltage propulsion that both are based upon the principles of electrogravitics as theoretically predicted by Dr. Paul LaViolette in his previously mentioned book, *SubQuantum Kinetics.*

The Hutchison Effect

In 1980, George Hathaway, a professional engineer licensed in Canada, along with entrepreneur, Alex Pizzaro, formed a small company to develop and promote what is referred to as "The Hutchison Effect." It is named after its inventor, John Hutchison, who liked to experiment with combinations of Tesla coils and Van de Graaff generators at the same time. Much of the information about the "lift and disruption" effects has been reported at various conferences (such as in the Third International Symposium on Non-Conventional Energy Technology held in Hull, Quebec in 1986). Videotapes of much of the phenomena have been shown on Japanese TV as well. Hathaway also assembled a three-hour videotape that documents the TV interviews, reports, and actual events. To summarize, the experiments were conducted with 250 kV of DC power on the Van de Graaff and about the same voltage of AC power on the Tesla coil. The total real power was about 1.5 KW continuously, according to Hathaway. Besides the disruptive effects, which were numerous, the lifting of various heavy objects by the field was most impressive. These events can be seen at the end of the commercial video, "Free Energy, The Race to Zero Point" (Lightworks AV, 1996) which is available for free on the internet. In regards to the AC contribution to the field, Hathaway reports that he measured a small voltage of 2 millivolts per meter in the active region (besides the DC offset). This is a small AC signal but on top of the high voltage DC signal, it performs amazing feats.

The importance of the Hutchison Effect to Brown electrogravitics is the AC "ripple" on the high DC voltage. A reference to this may be found in a military report by Dr. Dennis Cravens who gave T.T. Brown a high rating of "practicality." Cravens reported in his evaluation of Brown that older, high voltage supplies always had some AC ripple to the regulated signal, and wonders if this had any effect on Brown's phenomena [16] . Dr. LaViolette has also found this factor to have particular electrogravitic significance.

The B-2 Stealth Bomber Connection

Thanks to Dr. Paul LaViolette reporting in his article, "The U.S. Antigravity Squadron", there is substantial evidence that the electrogravitics research of the 1950's actually resulted in the B-2 Stealth Bomber "auxiliary propulsion system." Summarizing Dr. LaViolette's article, with references cited therein, the following facts are the most convincing:

- The B-2 charges the leading edges of its wing-like body, with high voltage;
- The B-2 is shaped just like T.T. Brown suggested an electrogravitic craft should look, for maximum charge separation;
- Northrup tested leading-edge charging in 1968;
- B-2 electrically charges the exhaust similar to the suggestion by T.T. Brown that the craft should be powered by needle-point flame-jet generators which electrically charge the exhaust;
- Aviation Week admits the existence of "dramatic, classified technologies" applicable to "aircraft control and propulsion" on the B-2;

- Aviation Week also disclosed that a high-density dielectric ceramic RAM made of powdered depleted uranium happens to have a mass density of 3 times that of the high-K dielectrics tested in the 1950's;
- The B-2's Emergency Power Units (EPU) can drive an electrical generator at high altitudes or even in space, since the fuel can be made to decompose rapidly even without oxygen;
- Edward Aldridge, former Secretary of the Air Force, admits that the B-2 creates no vapor trail at high altitudes. The decomposed gases from the EPU's can function as the ion-carrying medium, in much the same way as the hot exhaust gases from the air breathing flame jet generators.

These details create a sense of excitement about the world's foremost aircraft. Dr. LaViolette argues that the electrogravitic drive will function better at higher speeds due to the better flow of the ions. Therefore, it is likely, he says, that the B-2 actually is a supersonic aircraft, especially since the 1968 leading-edge charging experiments were for supersonic softening of the shock wave. House Representative Robert Walker was quoted recently in *Popular Science* ("Secrets of Groom Lake") as promoting the idea of declassifying military secrets that will help commercial development. We hope that this trend will continue so that advanced civilian spacecraft designs may also acquire an electrogravitic drive. Through a proposal he submitted in 1990, Dr. LaViolette made NASA aware that an electrogravitic drive would be a feasible propulsion method for the Mars journey because calculations show that the transit time can be less than one month, instead of half a year to a year. It is especially attractive since it uses so little power when it is operational and verified by Brown to work well in a vacuum. More information about electrogravitics and Brown's patents are available in my book, *Electrogravitics Systems: Reports on a New Propulsion Methodology*. Many of the books and videos mentioned in this article are available, as a public education service, from the non-profit organization, Integrity Research Institute, integrity-research.org.

4. Electrogravitics - Summary Description

Electrogravitics technology was being developed and tested in all major aerospace companies in the 1950s. Martin, Convair, Lear, Sperry, Raytheon, and many others all were eagerly studying electrogravitics. But in 1959 or early 1960, the technology became highly classified and the path to energy abundance was stymied.

A model theory that predicts electrogravitics, which some say is a far better one than relativity, is subquantum kinetics (SQK). Its developer, Dr. Paul A. LaViolette, started with chemical kinetics and came up with SQK after becoming very concerned by its gravitational predictions. He had not heard of the Biefeld-Brown Effect, nor the work of T. Townsend Brown. He struggled with these predictions until he encountered Brown's work, discovering that SQK predicted exactly what Brown was showing experimentally.

Why is SQK better than relativity? It takes no element on faith: Einstein's relativity is based on the faith that matter somehow "bends" space-time. Nowhere does he explain (or even try to) how or why this might be. One just has to take it on faith. It explains things without the need for "renormalization;" relativity ends in infinities without this mathematical fudge. It integrates EM with gravity. Einstein himself died still trying to do so. SQK also predicts things like the lensing of light, the apparent time dilation, and all other predictions of relativity, plus more -- all of which are testable (whereas string theory is 100% untestable).

In fact, many of the mysteries of Einsteinian mathematics are predicted in SQK.

Recommended reading is a book called *Secrets of Antigravity Propulsion*, by Dr. LaViolette, for more on the Biefeld-Brown Effect, Brown's work, electrogravitics, SQK, and black project technology.

In SQK, positively charged particles have a positive gravity potential "well." (Positive particles have positive gravity.) Negatively charged particles have a negative gravity potential "hill." (Negative particles have negative gravity.) The positive particle's "well" is just the smallest fraction bigger than the negative "hill," accounting for the apparent weakness of gravity, where essentially even amounts of positively charged and negatively charged particles make up common matter, and in quantities as big as the earth, say, offer a slight "well" overall, thus explaining why we are aware of gravity at all.

The Biefeld-Brown Effect was discovered when it was noted that a dielectric with electrodes attached at either end would apparently lose weight when charged if the positive pole was pointed upwards on a scale, and apparently gain weight when reversed, when placed on a scale. It was soon discovered that it was not a weight loss, but a movement in the direction of the positive pole. Higher voltage increased this effect. Further study showed that the higher the K of the dielectric, the more pronounced the movements became. Asymmetrical electrodes contribute to the increase of the effect, as well. Also, a non-linear dielectric produced greater movement than a linear one. With high K, non-linear dielectrics, a substantial force could be observed, even to the point of levitation with a local gravity field induced.

If, say, four devices were placed on a cross, each oriented 90° to the center and pointing in the same direction relative to that center, and a shaft was set at center and into a generator, when small amounts of energy are used at high voltage on the units, the resulting rotation will induce a greater output at the generator than the input to the units. Overunity, as energy is drawn from the zero point field, is achieved.

A petition for the release of electrogravitics from black projects has been started which demands this technology for civilian use: http://www.change.org/petitions/us-military-release-the-technology-of-electrogravitics [17].

5. NASA Finally Develops Electrogravitics

In 1994, IRI published the book, Electrogravitics Systems, Reports on a New Propulsion Methodology (IRI #611). It shows how popular T. T. Brown's inventions were in the 1950's and how Northrop applied the force-producing, high voltage, capacitor design to the B-2 bomber, lately called "field propulsion." Now, after eight years and thousands of books sold, NASA scientists claim they have discovered a "method for generating thrust from two dimensional asymmetrical capacitor modules," with their new US Patent #6,317,310. The results, they note, "are potentially greater efficiencies and improved reliability over currently available electric thrusters" (that expel gas). [18] (See NASA Press Release below).

The NASA promotion states, "In the past, inductive technology has been needed to create thrust, rotational motion, or step an actuator using electricity. This new technology accomplishes these tasks and more by using high potential, low current asymmetrical capacitor modules…It had been theorized that thrust generation from this phenomenon was feasible, but no working prototypes had been developed, until now." While historically, based on all of Brown's research, the last sentence is obviously incorrect, this author is gratified that fuelless propulsion is finally being utilized by NASA. JLN Labs have test results of their prototype of the thruster at http://jnaudin.free.fr/html/omptv1.htm while Gravitec, Inc. also has test results from Purdue University at http://foldedspace.com/corporate.html with Brown's term "Electrokinetic Propulsion."[19] (Credit: Thomas Valone, FUTURE ENERGY, VOL. 1, NO. 3, 2002).

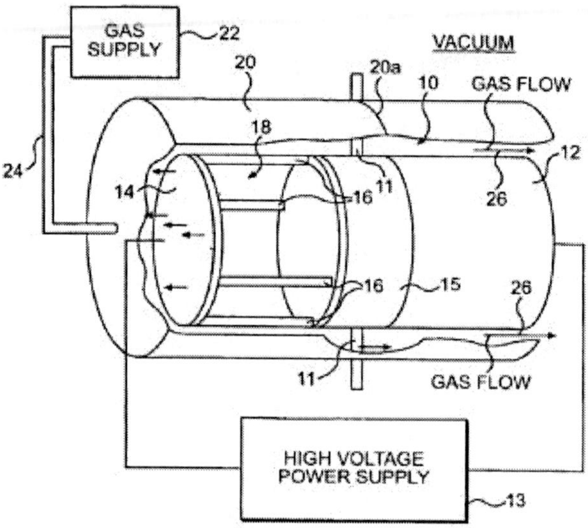

Figure 6. Apparatus and Method for Generating Thrust Using a Two Dimensional Asymmetrical Capacitor Module.

NASA Press Release, 11/6/01 "Apparatus and Method for Generating Thrust Using a Two Dimensional Asymmetrical Capacitor Module" Objective: NASA scientists have discovered a method for generating thrust from two dimensional asymmetrical capacitor modules. The results are potentially greater efficiencies and improved reliability over currently available electric thrusters.

Description: In the past, inductive technology has been needed to create thrust, rotational motion, or step an actuator using electricity. This new technology accomplishes these tasks and more by using high potential, low current asymmetrical capacitor modules. The dielectric material of a capacitor under high voltage experiences a force. Based on the geometry of the capacitor, its material properties, and ambient conditions, the force can be predicted and utilized to move the entire capacitor and its mounting in a predictable direction. It had been theorized that thrust generation from this phenomenon was feasible, but no working prototypes had been developed, until now. Benefits - Improved reliability - Potential use of atmosphere as propulsive medium - Reduced cost of orbit Low current draw Current Capabilities This MSFC invention, the Two Dimensional Asymmetrical Thrust Capacitor, has been demonstrated in with a proof-of-concept model. Although additional experimental optimization is indicated, this technology will provide improved reliability due to the use of no moving parts to create thrust in a gaseous environment. Thrust is expected in aqueous environments as well. Potential Application(s) - Linear accelerator to launch payloads - Rotate a propeller or other shaft driven application - Stepping actuator - Near-earth orbital maneuvering Technology Commercialization Status Protection - NASA has filed a patent application for this technology. Prototype - A proof-of-principal model is available. Licensing - Licenses are available for all fields of use. Knowledge Transfer - Inventor is available for assistance to license (see [20, 21]).

CONCLUSION

Thus we see from the innocent experiments of Professor Nipher, an evolutionary wave of development occurred in the past century to usher in a new form of propulsion, which can operate without burning fossil fuels. Today, electrogravitics and electrokinetics still remain unknown to most scientists, except for the engineers in black projects. More about this is in the last chapter of this book.

REFERENCES

[1] Atlantis Rising, Number 22, p.35; *AIR International*, Jan., 2000; Jane's Defence Weekly, 10 June 1995, p.34.

[2] Electrogravitics Systems: Reports on a New Propulsion Methodology, Integrity Research Institute, 1994, edited by Thomas Valone.

[3] Nipher, "Can Electricity Destroy Gravitation", *Electrical Experimenter*, March, 1918.

[4] *Trans. Acad. Sci.* St. Louis, pp. 163-175 and 177-193 (July 1916).

[5] *Trans. Acad. Sci.* of St. Louis XXVII: 383-387 (March 2, 1920).

[6] *Proc. Amer. Phil. Soc.* Philadelphia 52: 283-6; Science (Sept. 1, 1911), pp. 282-3.

[7] *Experimental Studies in Electricity and Magnetism*, pp. 19-24.

[8] Nipher: "A Flash of Lightning", *Popular Science Monthly*, Jan. 1912: p. 76.

[9] Electrogravitics II: Validating Reports on a New Propulsion Methodology (edited by Thomas Valone, Integrity Research Institute, 2000, p. 71.

[10] Causality, *Electromagnetic Induction and Gravitation*, Electret Scientific Co., POB 4132, Star City WV 26504, p. 29-31.

[11] Paul LaViolette, Starburst Foundation, May 13, 2012, http://starburstfound.org/electrogravitics blog/?cat=4.

[12] Secrets of Antigravity Propulsion, by Paul LaViolette (Bear and Co., 2008).

[13] Towards flight without stress or strain or weight, *Interavia Magazine*.

[14] "*Military Power*" published in a British aviation magazine, AIR International, (Jan., 2000.

[15] Antigravity: The Dream Made Reality book [self-published by John Thomas.

[16] Cravens, T.L. "*Electric Propulsion Study*", AL-TR-89-040, #ADA 227121, Science Applic. Inter. Corp., Torrance, CA 90501

[17] Amaterasu Solar, April 1, 2012, http://www.activistpost.com/2012/04/ electrogravitics-simplified-description.html

[18] See NASA TechFinder: http://technology.nasa.gov/scripts/nls_ax.dll/ twDispTOPSItem (111;TOP8-80;0;1) for this information. Surprisingly however, no mention of the famous T. Townsend Brown patents (e.g. US #2,949,550 or #3,022,430), the Biefeld-Brown Effect, or the terms "electrogravitics" or "electrokinetics" is made in the technical literature or in the NASA patent itself.

[19] Thomas Valone, *Future Energy*, VOL. 1, NO. 3, 2002.

[20] Contact If your company is interested in this technology please contact: Sammy Nabors NASA Marshall Space Flight Center M.S. CD30 George C. Marshall Space Flight Center MSFC, AL 35812 Phone: (256) 544-5226 FAX: (256) 544-3151

E-mail: sammy.nabors@msfc.nasa.gov

Related Technologies TWO DIMENSIONAL, ASSYMMETRICAL CAPACITOR MODULES UNDER HIGH POTENTIAL FOR GENERATING THRUST.

[21] Public Release Date 11/6/01

THE UNITED STATES GOVERNMENT DOES NOT ASSUME ANY LIABILITY RESULTING FROM THE USE OF THE INFORMATION PROVIDED BY THIS SERVICE AND DOES NOT REPRESENT OR

WARRANT THAT USE OF THE INFORMATION WILL BE FREE FROM PRIVATELY OWNED RIGHTS. IN ADDITION, THE UNITED STATES GOVERNMENT DOES NOT ENDORSE ANY COMMERCIAL PRODUCT, PROCESS, OR ACTIVITY IDENTIFIED BY THIS SERVICE.

Chapter 2

ON THE POSSIBILITY OF STRONG COUPLING BETWEEN ELECTRICITY AND GRAVITATION

*Takaaki Musha**

Advanced Sci.-Tech. Research Organization, Yokohama, Japan

ABSTRACT

T.T.Brown, who discovered the electrogravitic effect, performedseveral experiments during 1950s that succeeded to generate thrust without the reliance on a surrounding medium by applying high voltages to materials with high dielectric constants. In the late 1980 and early 1990s, a rash of observer "sightings" of unidentified high speed, high flying air vehicles would seem to indicate the possible application of this effect.. In the later 1990's BAe also provided internal resources for its own anti-gravity studies known collectively as "Project Greenglow." In Thomas Valone's book, "Electrogravitic Systems", published in 1994, P.A.LaViolette claimed that electrogravitic technology was developed under U.S. Air Force black project since late 1954, and that it may now have been put to practical use in the B-2 Advanced Technology Bomber to provide an exotic auxiliary mode of propulsion. The supposed electrogravitic drive of B-2 could allegedly allow it to fly at a sufficiently high speed and high altitude to even reach the edge of space. Subsequently there has been no significant progress in this area, but late 1990s, the group of the Honda Corporation Research Institute and the author conducted a series of experiments to confirm the electrogravitic

* E-mail address: takaaki.mushya@gmail.com.

effect and obtained positive results. In 2004, Musha and Ivanov independently published papers on the electrogravitic effect for a dielectric material under high voltage, demonstrating a connection between electromagnetic and gravitational forces. This chapter outlines the physical basis of electro-gravitics and related physics which might lead to thc development of advanced technologies for future aviation systems.

1. INTRODUCTION

Prof.Biefeld and T.T.Brown discovered that a high potential charged capacitor with dielectrics exhibited unidirectional thrust toward the positive plate when the atoms of a material are placed within the electric field of a capacitor [1]. This phenomenon is called the Biefeld-Brown effect (B-B effect) and it suggests a connection between electricity and gravitation.

Characteristics of the B-B effect can be summarized shown as follows[2]

1. The separation of the plates of the condenser—closer plates, greater effect.
2. The higher the specific inductive capacity of the dielectrics between the plates, the greater the effect.
3. The greater the area of the condenser plates, the greater the effect.
4. The greater the voltage differences between the plates, the greater the effect.
5. The greater the mass of the material between the plates, the greater the effect.

However, the coupling between electrostatic and gravitational fields can be predicted neither by General Relativity, nor by conventional field theory. The author attempts to explain this phenomenon by introducing a new gravitational field generated inside the atom by a high potential electric field.

The Biefeld-Brown effect is known to encompass electrokinetic effects, but is often claimed to have an electrogravitic component as well. The latter was reported by Thomas Townsend Brown to generate thrust without the reliance on a surrounding medium (e.g., air) by applying high voltages to materials with high dielectric constants. Correlations between variations in thrust and the position of the Moon led Brown to associating the electrostatic phenomenon with gravity. Thomas Townsend Brown's first article and patent emphasized electrogravitics.

Recent electrogravitic tests in Canada and Japan have been reportedly positive. Aviation photographer, Doyle R. Buehler, and Takaaki Musha, member of the Advanced Space Propulsion Investigation Committee under the Japan Society for Aeronautical and Space Sciences reported the positive results. Buehler conducted his experiments at the Southern Alberta Institute of Technology, in Calgary, Canada [3]. Musha conducted experiments at the Honda Corporation Research Institute from the first of February to the first of March of 1996. [4]

Figure 1. Giant lifter experiment at Meisei University, Japan.

In 2003, the group of Meisei University also conducted the experiment by manufacturing a giant lifter and they succeeded for its uplifting to the air with a stable hovering at 15 meters above the ground inside the gymnasium as shown in Figure 1. The voltage used during this experiment was 41kV DC and the electric current was 10.7 mA [5]. A research group at Honda under Okamoto replicated his experiments with higher voltages and similar positive results [4]. Musha used a weak field approximation of Einstein's General Relativity Theory to generate a formula to explain electrogravitics;

$$E_g \approx -Z\sqrt{4\pi\varepsilon_r\varepsilon_0 G}\cdot E = -8.62\times 10^{-11} Z\sqrt{\varepsilon_r}\cdot E \ ,$$

where Z is a number of electrons circulating around the atomic nucleus, ε_r is a specific inductive capacity of the dielectric material, ε_0 is a permittivity of free space and G is the gravitational constant.

Analyses by Noriki Iwanaga of Musha's theoretical explanations were judged to be flawed from the standpoint of general relativistic side, but he also found that in lieu of a better theory it would be profitable to apply the Brown's propulsion method for small space vehicles as if the Musha's theory was physically valid[6].

Five years later, a firm theoretical foundation for Brown's electrogravitic effect was put forth by Boyko Ivanov [7] of the Institute for Nuclear Research and Nuclear Energy, Sofia, Bulgaria. He used classical approaches to Einstein's equations known as the Weyl-Majumdar-Papapetrou field solutions, dating back to 1916, to derive what he called root gravity. Ivanov's initial proofs were released through the Los Alamos National Laboratory archives [8, 9]. They contained descriptions of eight types of experiments that could be performed in the laboratory to detect root gravity. The first one incorporates parallel plate capacitors carrying several hundred thousand volts. His formula for the electrostatic production of root gravity supports reports by Brown, Buehler, and Musha. Invanov's and Musha's formulas are equivalent when the dielectric atomic number Z equals unity [4]. Both are inverse functions of the dielectric thickness and directly proportional to the voltage. In 2004, Musha incorporated modulo Z and international system units (SI) and Ivanov used the CGS (Gauss) system of units. Because both of them are linear approximation for the case of almost flat metric, neither formula accounted for the correlations between the electrogravitic variations of the Biefeld-Brown effect and the positions of the Moon that had been reported by Brown and Musha.

Statistical analyses of the electrokinetic variations of the Biefeld-Brown effect yielded insignificant correlations with lunar positions. It was reported by G.V.Stephenson [10] that measured Biefeld-Brown effect showed a dependence on thunderstorm activity. Thus neither formula is applicable to the electrokinetic aspect of the Biefeld-Brown effect.

2. Theoretical Considerration on the B-B Effect

Weak field approximation of Einstein's general relativity leads to the generalized formula of Lorenz force given by [11];

$$F = q(E + v \times B) + m(E_g + v \times B_g), \tag{1}$$

where q is charge of the particle, m is mass of the particle, E is electric field, B is magnetic field, v is velocity of the particle, E_g is electrogravitic field and B_g is magnetogravitic field.

From this equation, it can be seen that there exists a gravito-electromagnetic forces to generate electric and gravitoelectric coupling.

It was predicted by Schiff and Barnhill [12] that there exists net forces due to interplay between gravity and electrostatic forces inside a metallic shell, that for no current, can be shown as[13]

$$qE + mE_g = 0, \tag{2}$$

in the nonrelativistic limit. From which we can suppose that a new gravitational field can be generated inside a dielectric material, which has no current, in order to prevent drift of atomic nuclei due to an external electric field. Then there will be an electrically induced gravitational field generated at the center of the charged particle by an external electric field given by[14]

$$E_g = -(q/m)E, \tag{3}$$

which induce the change of the electrogravitic potential around the atom shown as follows:

Comparing q / m values of an electron and a pion, E_g can be generated by an electron rather than a pion, hence we can let $q \approx e$ and $m \approx m_e$ where e is a charge of an electron and m_e is its mass. For the estimation of gravitational cffect, we introduce the following approximation of the electrogravitic potential given by

$$\Phi_g = -\frac{e}{m_e} E\{\delta^2 x / (\delta^2 + x^2)\}, \tag{4}$$

which satisfies following conditions:

$$\frac{\partial}{\partial x} \Phi_g(0) = -\frac{e}{m_e} E, \tag{5.1}$$

$$\frac{\partial}{\partial x} \Phi_g(\pm\infty) = 0, \tag{5.2}$$

where δ is a length of the domain at which the new gravitational field is generated.

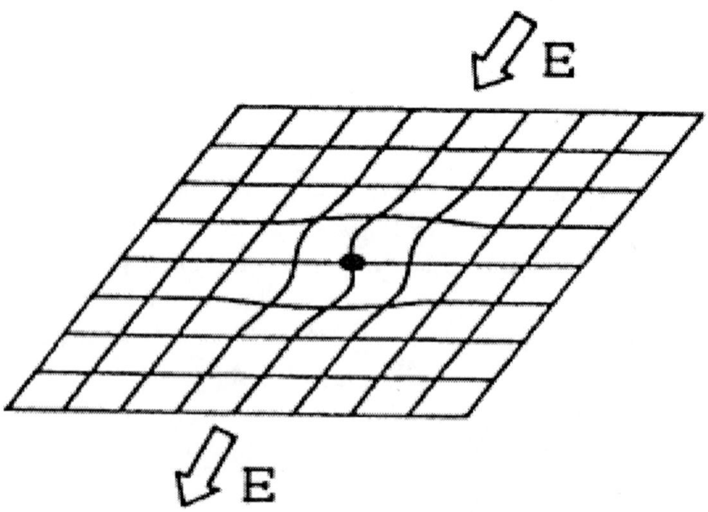

Figure 2. New gravitational field generated by an external electric field.

Figure 2 shows the deformation of the space generated at the center of the elementary particle and the shape of the electrogravitic potential orbital radius of the electron around the nucleus.

By the asymmetric electron orbit shown in Fig 3 generated by the external electric field, the electrogravitic potential at the center of the atomic nucleus becomes

$$\Phi_g = \Phi_g(r+\lambda) + \Phi_g(r-\lambda), \tag{6}$$

where λ is a displacement of charge by an applied electric field and r is the radius from which the electrogravitic force generated by electrons circulating around the nucleus. The number of them equals the atomic number Z, which is given by [15,16]

$$E_g = \frac{\partial}{\partial x}\Phi_g = \delta^2 \frac{e}{m_e} E \sum_{i=1}^{Z} [(r_i+\lambda)^{-2} + (r_i-\lambda)^{-2}] \ , \tag{7}$$

when satisfying $|r_i \pm \lambda| >> \delta$, where r_i is the orbital radius of each electron around the atomic nucleus.

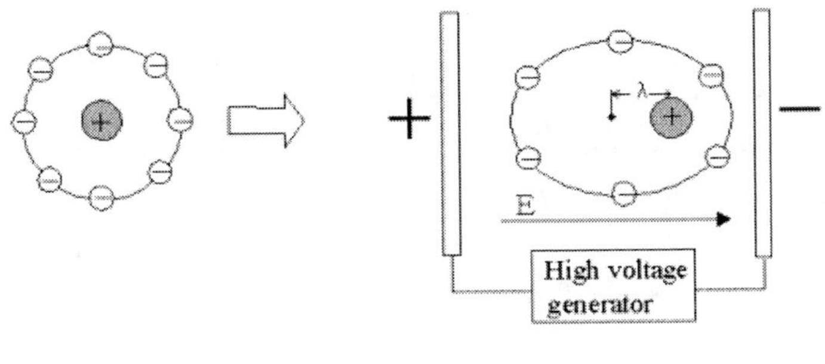

Figure 3. Deformation of the atom in an external electric field.

For simplifying the problem, we set $r_i \approx r_0$ (r_0: r Bohr's radius), then Eq.(7) becomes

$$E_g \approx Z\delta^2 \frac{e}{m_e} E[(r_0+\lambda)^{-2} + (r_0-\lambda)^{-2}] \ . \tag{8}$$

For relative lower voltage, we can suppose that $r_0 >> \lambda$. Then Eq. (8) can be approximated as

$$E_g \approx -2\delta^2 Z \frac{e}{m_e r_0^2}\{1 + 3\lambda^2 / r_0^2 + \cdots\}E. \tag{9}$$

If we suppose that additional equivalent mass in a space due to the electric field [17] is canceled by negative mass created by the new gravitational field (derivation of which is shown in Appendix), we have the following formula given by

$$\int_V 2\pi G\varepsilon E^2 / c^2 dv - \int_V E_g^2 /(2c^2)dv = 0 , \tag{10}$$

where ε is permittivity, G is a gravitational constant and c is a light speed.

From which, we have

$$\delta \approx \sqrt[4]{\pi\varepsilon G m_e^2 r_0^4 / e^2} . \tag{11}$$

If ε equals ε_0, which is the permittivity of free space, we obtain the length of a domain where the new gravitational field is generated to be $\delta \approx 8.3 \times 10^{-22}$ (m), that is much smaller than the radius of electron, which is about 2.8×10^{-15} (m).

Then acceleration of the dielectric material induced high potential electric field can be approximated from Eq.(9) and (11) as

$$E_g \approx -Z\sqrt{4\pi\varepsilon_r\varepsilon_0 G} \cdot E = -8.62 \times 10^{-11} Z\sqrt{\varepsilon_r} \cdot E, \tag{12}$$

where ε_r is specific inductive capacity of the dielectric material determined as $\varepsilon = \varepsilon_r \varepsilon_0$.

From which, the force generated by high potential electric field becomes

$$F \approx 8.62 \times 10^{-11} Z\mu S\sqrt{\varepsilon_r} \cdot V / t, \tag{13}$$

where μ is the total mass of the dielectric material per unit area, S is the area of the capacitor, t is the thickness of the capacitor and V is the applied voltage.

This equation presented here satisfies characteristics (1) ~ (5) mentioned in the previous section and it also predicts another characteristic for the B-B effect shown as follows.

(1) The greater the atomic number of the material between the plates, the greater the effect. This should be straightforward to prove through experimental means.

3. Relation with Weyl Electrovacuum Solutions Derived by Ivanov

Another theoretical foundation for Brown's electrogravitic effect was put forth by Boyko Ivanov of the Institute for Nuclear Research and Nuclear Energy, Sofia, Bulgaria.

He used classical approaches to Einstein's equations known as the Weyl-Majumdar-Papapetrou field solutions, dating back to 1916, to derive what he called root gravity [7-9].

In general relativity, EM fields do indeed alter the metric of space-time and induce a gravitational force through their energy-moment tensor given by

$$T_{\nu}^{\mu} = \frac{1}{4\pi}\left(F^{\mu\alpha} F_{\nu\alpha} - \frac{1}{4}\delta_{\nu}^{\mu} F^{\alpha\beta} F_{\alpha\beta} \right), \tag{14}$$

where $F_{\mu\nu} = \partial_{\mu} A_{\nu} - \partial_{\nu} A_{\mu}$ is the electromagnetic tensor and A_{μ} is the four potential.

Ivanov had taken into account that $T_{\mu}^{\mu} = 0$ and he further assumed that the metric and E-M fields do not depend on time. In this stationary case, the problem can be simplified by setting $A_{\mu} = (\overline{\phi},0,0,0)$, then there is just an electric field given by

$$E_{\mu} = F_{0\mu} = -\overline{\phi}_{\mu}, \tag{15}$$

Let us further assume that the space-time is static, so that $f \equiv g_{00} = F(\phi)$ has the unique form $f = 1 + B\phi + \phi^2$, which was found by Weyl in 1917 in the axially-symmetric case, solutions of which are known as Weyl fields. Then the equation for the gravitational field induced by static electric field can be given by

$$g_i = c^2 f^{-1}\left(\frac{B'}{2}\sqrt{\frac{\kappa\varepsilon}{8\pi}}\overline{\phi}_i + \frac{\kappa\varepsilon}{8\pi}\overline{\phi}\,\overline{\phi}_i\right), \tag{16}$$

where $f \equiv g_{00}$, B' is a constant and $\kappa = 8\pi G / c^4$.

From this, Ivanov derived the formula of gravitational force F_g shown here for a capacitor as in [7]

$$F_g = \sqrt{G\varepsilon}\frac{M}{d}\overline{\psi}_2 = \sqrt{G\varepsilon}\mu S\overline{\psi}_2, \tag{17}$$

where M is the mass of the dielectric, μ is its mass density, ε is the dielectric constant, d is the distance between the plates, ψ_2 is the potential of the second plate when $\psi_1 = 0$, and S is an area of the plate.

Ivanov also had shown that magnetic fields would produce the same effects as electric ones. From the definition of the magnetic field given by

$$H_i = -\frac{1}{2}\sqrt{-g}\varepsilon_{ikl}F^{kl}, \tag{18}$$

Analogous to those for electric fields, Ivanov obtained the formula for the connection between magnetic and gravitational fields given by

$$g_i = -\sqrt{G\mu}H_i, \tag{19}$$

where μ denotes the magnetic constant.

As mentioned in Ivanov's correspondence to the author, Invanov's and the author's formulas are equivalent when the dielectric atomic number Z equals unity. Both are inverse functions of the dielectric thickness and directly proportional to the voltage. The author incorporated modulo Z and international system units (SI) and Ivanov used the CGS (Gauss) system of units. From which, it is seen that the derived equation shows the validity of electrogravitic effect as claimed by T.T.Brown.

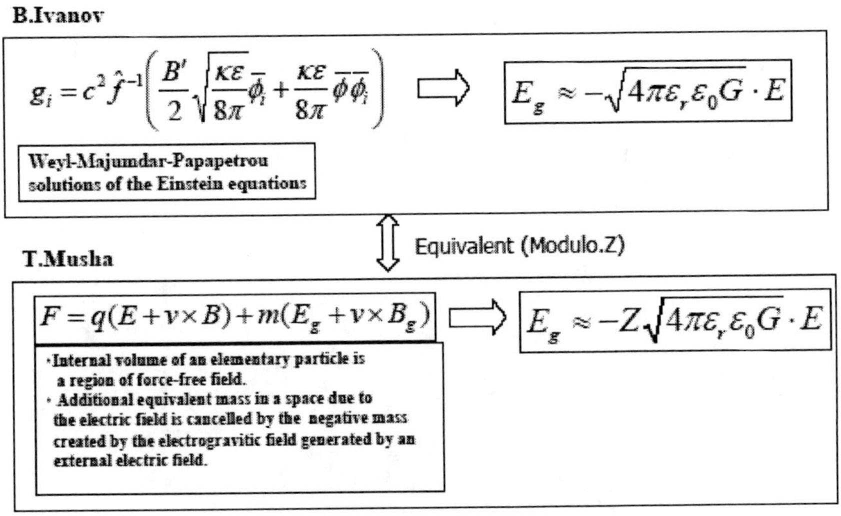

Figure 4. Correspondence between Ivanov theory and the Musha's formula.

4. Experimental Results and Their Analysis

From the 1st of Feb. till the 1st of March in 1996, the research group of the HONDA RandD Institute conducted experiments to verify the B-B effect with an improved experimental device [18,19] to reject the influence of corona discharges and electric wind around the capacitor. For the rejection of these effects, the capacitor was set in the insulator oil contained within a metallic vessel as shown in Figure 5.

The capacitor used at the experiment was a circular plate made of glass with thickness=1mm, diameter=170mm and weight=62g. The specific inductive capacity of the capacitor was about 10.

They conducted experiments for both cases, +DC 18kv and -DC 18kv applied to the capacitor. The experimental results obtained by the HONDA research group is shown in Figure 6, where the horizontal line is for the number of the experiment which they conducted and the vertical line is for the weight loss of the capacitor.

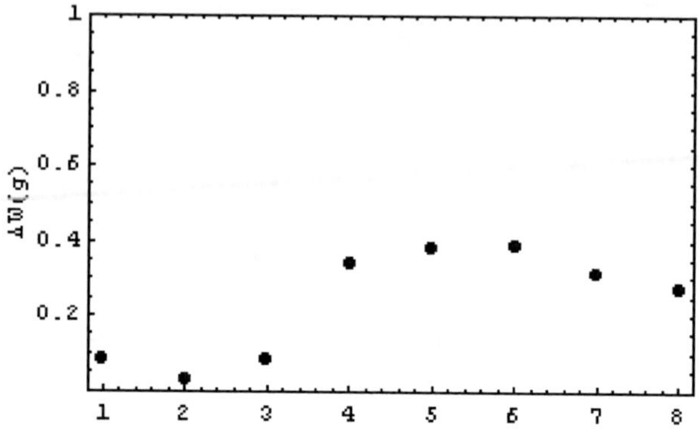

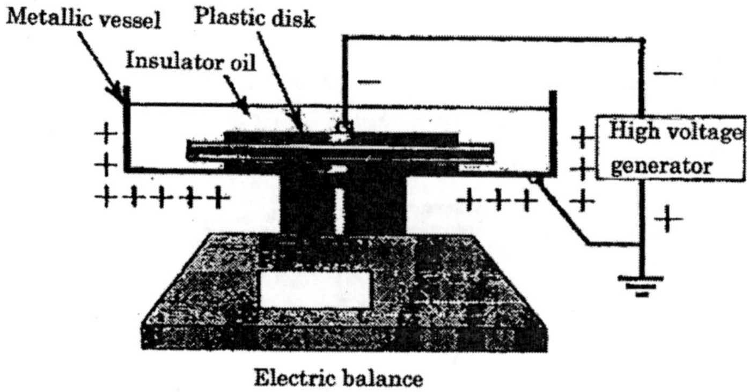

Figure 5. Experimental set-up conducted by Honda research group.

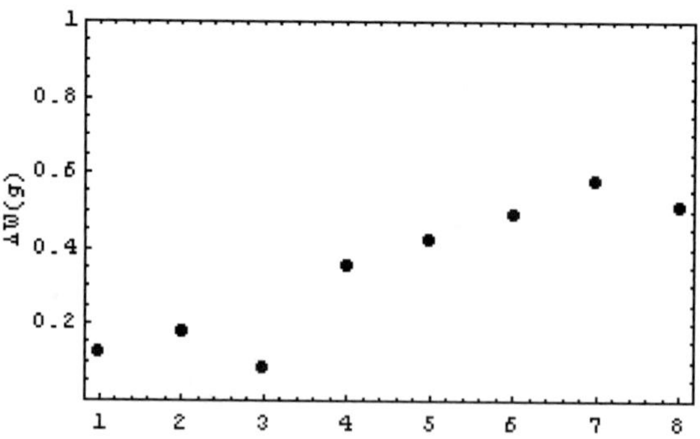

Figure 6. Weight reduction observed at the experiment (left figure: +18kV, right figure: -18kV).

By the experiment results, the coefficient κ as follows can be determined by

$$\kappa = E_g / E = (\Delta W / W)(t / V) g_0 \ , \tag{20}$$

where ΔW is a weight loss of the capacitor under high potential electric field, W is a weight of the capacitor and g_0 is a gravitational acceleration on the Earth.

A histogram of the experiment was obtained as shown in Figure 7, where the horizontal line shows the value K, which equals κ times 10^9, and the vertical line shows the number of observed results. Statistical analysis shows that the mean value of K obtained is 2.59 and its standard deviation is 1.46, then we have $\kappa \approx (2.59 \pm 1.46) \times 10^{-9}$.

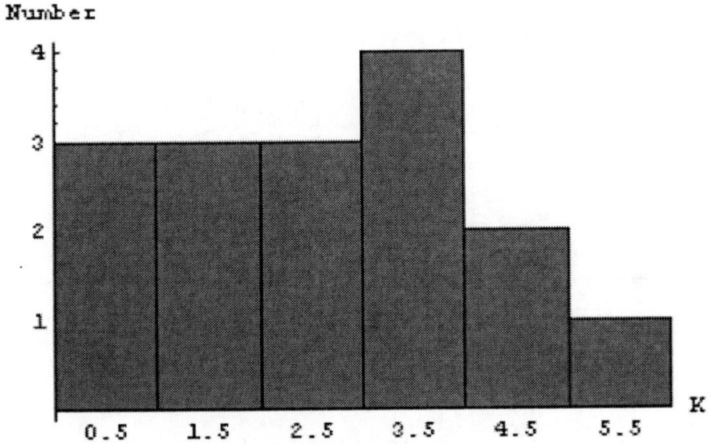

Figure 7. Histogram of the experimental results.

From Eq.(12), the absolute value of κ can be given by

$$|\kappa| = 8.62 \times 10^{-11} Z \sqrt{\varepsilon_r} \ . \tag{21}$$

Most of ingredient of glass is SiO2 , the mean value of Z per unit atom can be roughly estimated as $Z \approx (14 + 8 \times 2)/3 = 10$, so we have $|\kappa| \approx 2.73 \times 10^{-9}$. From which, we have

$$(E_g)_{measured} / (E_g)_{theoretica\,l} = 0.95 \pm 0.53 \ . \tag{22}$$

Thus it is considered that the force generated due to the external electric field can be a new gravitational field generated in the micro level by strong coupling between electromagnetism and gravitation.

5. Applications to Human Space Flight

By using the equation which the author derived, Iwanaga analyzed the effectiveness of the B-B effect as a propulsion system in his paper[6] and he estimated the thrust obtained by the B-B effect is much better than that of arc-jet propulsion system as shown in Table 1. He considered it would be profitable to apply this propulsion method to small space vehicles.

Table 1. Thrust of the propulsion systems and B-B effect

Propulsion System	Thrust (N)
Jet engine	2×10^{5}
Chemical fuel rocket	2.45×10^{5}
Arc-jet	1.50×10^{-1}
Nuclear power	8.82×10^{5}
Photon rocket	3.3
B-B effect*	100

*Calculated for mass=100Kg, diameter/height ratio=100, $\varepsilon_r = 5$ and $E = 7 \times 10^{8}$ V/m[6].

6. Gravity Induced by Electric Charge for Stars

Most of the space within our galaxy is occupied by plasma (i.e a rarefied ionized gas) containing electrons (negative charges) and ionized atoms (positive charges). Every charged particle in the plasma has an electric potential energy (voltage) just as every pebble on a mountain has a mechanical potential energy with respect to sea level. The Sun is surrounded by a plasma cell that stretches far out - many times the radius of Pluto [20].

The Sun is at a more positive electrical potential (voltage) than is the space plasma surrounding it - probably in the order of 10 billion volts. Positive ions leave the Sun and electrons enter the Sun. Both of these flows add to form a net positive current leaving the Sun as shown in Figure 8. This constitutes a plasma discharge analogous in every way to those that have been observed in electrical plasma laboratories for decades. Because of the Sun's positive charge (voltage), it acts as the anode in a plasma discharge. As such, it exhibits many of the phenomena observed in earthbound plasma experiments, such as anode tufting. The granules observed on the surface of the photosphere are anode tufts (plasma in the arc mode).

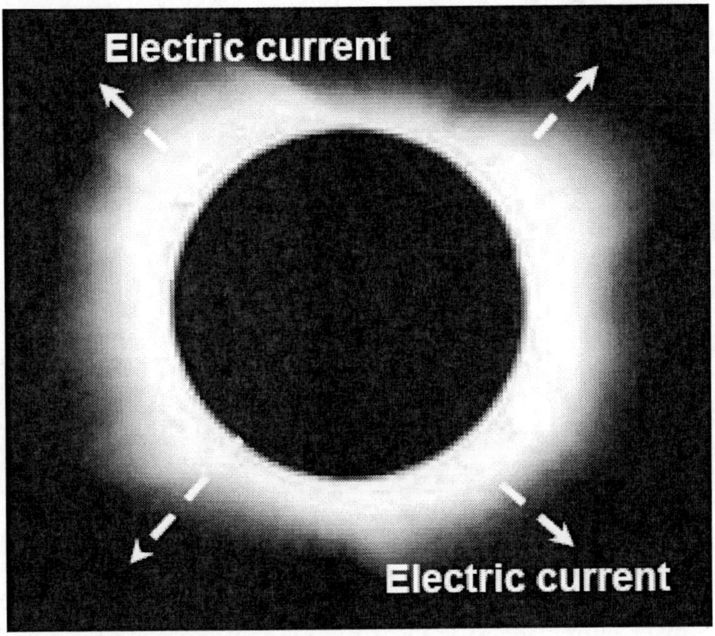

Figure 8. Electric current generated around the Sun.

The Sun may be powered, not from within itself, but from outside, by the electric (Birkeland) currents that flow in our arm of our galaxy as they do in all galaxies as shown in Figure 9.

This possibility that the Sun may be externally powered by its galactic environment is the most speculative idea in the ES (Electric Sun) hypothesis [21] and is always attacked by critics while they ignore all the other explanatory properties of the ES model. In the Plasma Universe model, these cosmic sized, low-density currents create the galaxies and the stars within those galaxies by the electromagnetic z-pinch effect. It is only a small extrapolation to ask whether these currents remain to power those stars. Galactic currents are of low current density, but, because the sizes of the stars are large, the total current is high. The Sun's radiated power at any instant is due to the energy imparted by that amperage.

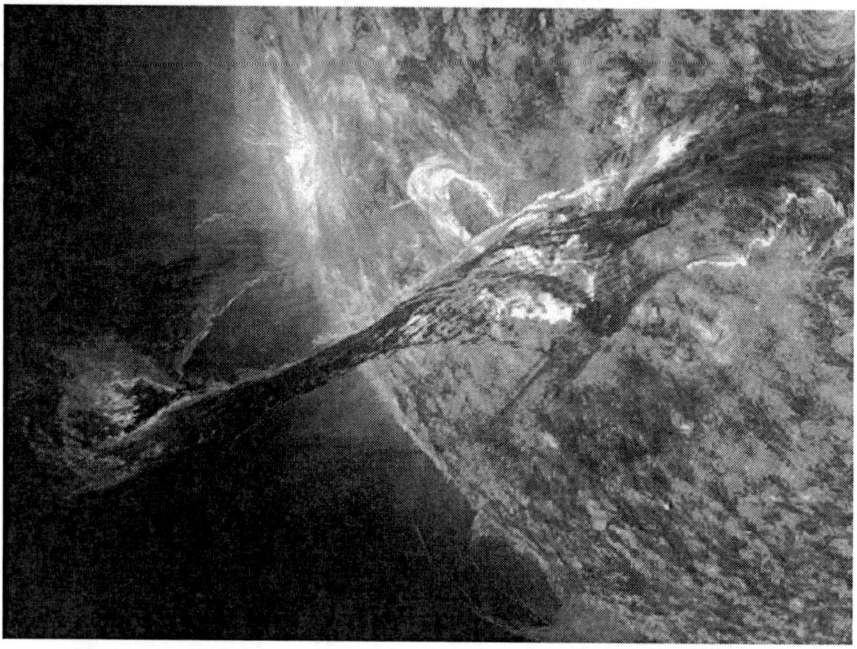

Figure 9. Solar flare showing the twisting motion characteristics of a Birkeland current.

As the Sun moves around the galactic center it may come into regions of higher or lower current density and so its output may vary both periodically and randomly. If the electrogravitic equation given by Eq.(6) can be applied to massive stars, it is considered that they may carry electric charges.

The hypothesis that the Sun may have an electric charge given by $Q_S = 5 \times 10^{18}$ (C), was proposed by Prof.Bailey in 1960 for the explanation of the maximum energy found for a primary cosmic ray particle and other astronomical phenomena [22]. He considered that the four dimensional space-time universe was a hyper-surface in a five dimensional universe and there existed streams of electrically charged particle from hyper dimensional universe into four dimensional space-time, where the laws of conservation of energy momentum and electric charge held true. He also thought that these electrical streams depends on local metric of space-time.

From the equation in the paper, "On the gravitational field induced by static electromagnetic source"[8] or the equation in the paper, " Strong gravitational force induced by static electromagnetic fields"[9], the electrogravitic field can be given by

$$g(r) = -\sqrt{4\pi\varepsilon_0 G}\,\frac{\overline{\psi}\cdot r_0}{r^2}, \tag{23}$$

where $\overline{\psi} = \dfrac{Q}{4\pi\varepsilon_0 r_0}$ (Q: charge of the sphere).

From the formula, $div\ g = 4\pi G\rho_m$, and the Gauss's theorem, $\int_S div\ g\cdot ds = \int_V 4\pi G\rho_m dv$, where ρ_m is the mass density inside the sphere, we have

$$-\frac{\sqrt{4\pi\varepsilon_0 G}}{4\pi\varepsilon_0 r_0^2} Q\int_S ds = 4\pi G M_0 , \tag{24}$$

where $M_e = \int_V \rho_m dv$.

Finally we obtain the equivalent mass generated by the electric charge becomes

$$M_e = -\frac{Q}{\sqrt{4\pi\varepsilon_0 G}}. \tag{25}$$

This equation shows the gravitational effect of electric field around the spherical body. When we suppose that the Sun has the net negative charge, $Q_S = 5\times 10^{18}$ C, the equivalent mass generated by this electric field can be estimated by Eq.(25) as $M = 1.74\times 10^{28}$ (kg), which is about one percent of the mass of our Sun.

From this calculation result, it is considered that some of the gravitational field attributes to the electric charge of the Sun.

Bailey proposed a formula for a star like the Sun of the mass M_{S_S} (kg) which carries a net negative charge, $-Q_S$ (C) , given by[20]

$$Q_S = \beta\sqrt{4\pi\varepsilon_0 G}M_S, \tag{26}$$

where β is a pure number of the order of 0.03.

By introducing Eq.(25) into Eq.(24), we have $M_e = \beta M_S$, which means some ratio of the gravitational mass of a star like the Sun is eletrogravitic origin.

CONCLUSION

The theoretical equation for the B-B effect, which satisfies the characteristics clarified T.T.Brown, is derived. As the experimental results agrees well with the theoretical calculation, it is considered that the force generated due to the external electric field can be due to a new gravitational field generated in the microscopic level induced by strong coupling between electromagnetism and gravitation.

APPENDIX

According to the book, "Causality Electromagnetic Induction and Gravitation", by O.D. Jefimenko [23], we have the equation;

$$\nabla \cdot E_g = -4\pi G\rho + \frac{E_g^2}{2c^2}, \tag{A.1}$$

where E_g is a gravitational field.

From the assumption that the mass density due to the electromagnetic field equals the negative mass density generated by the new gravitational field due to the external electric field. The mass density due to the external field becomes $\rho = \varepsilon E^2 / 2$, according to the electromagnetic theory. By introducing this formula into Eq.(A.1), we have

$$\nabla \cdot E_g = -\frac{2\pi G\varepsilon}{c^2} E^2 + \frac{E_g^2}{2c^2}. \tag{A.2}$$

From my assumption, the equivalent mass due to the electric field impressed to the dielectric material and the equivalent mass generated by the new gravitational field cancel each other, then we obtain

$$\int_V 2\pi G\varepsilon E^2 / c^2 dv - \int_V E_g^2 /(2c^2) dv = 0 . \qquad \text{(A.3)}$$

REFERENCES

[1] INTEL, Toward Flight without Stress or Strain … or Weight, Interavia, XI;1956, 373-374.

[2] Sigma, Rho, Eher-Technology: A rational approach to gravity control, Cadake Industries, *Clayton*, 1986.

[3] Bueher,D.R., Exploratory research on the Phenomenon of the Movement of High Voltage Capacitors, *Journal of Space Mixing*, 2; 2004, 1-22 .

[4] T.Musha, T., Theoretical Explanation of the Biefeld-Brown Effect, *Electric Spacecraft Journal,* Issue 31; 2000, 29.

[5] Musha,T., Connection Between Einstein's Unified Field Theory and the Biefeld-Brown Effect, *Infinite Energy,* Issue 104; 2012, 27-30.

[6] Iwanaga ,N., Review of Some Field Propulsion Methods from the General Relativistic Standpoint, Space Technology and Applications International Forum, *The American Institute of Physics*,1999, 1051-1059.

[7] Ivanov, B.V., Weyl Electrovacuum Solutions and Gauge Invariance (Preliminary research for gravity control), *Journal of Space Exploration*, Vol.2 (1); 2013, 49-51.

[8] Ivanov, B.V., Strong gravitational force induced by static electromagnetic fields, arXiv: gr-qc/0407048 v1, July 13, 2004.

[9] Ivanov, B.V., On the gravitational field induced by static electromagnetic sources, arXiv: gr-qc/0502047 v1, February 10, 2005.

[10] Stephenson, G.V., *The Biefeld Brown Effect and the Global Electric Circuit,* STAIF, 2005.

[11] Harris,E.G., Analogy between general relativity and electromagnetism for slowly moving particles in weak gravitational fields, *American Journal of Physics*, 59(5); 1991, 421-425.

[12] L.I. Schiff, M.V. Barnhill, Gravitation-Induced Electric Field Near a Metal, *Physical Review,* 151(4); 1966, 1067-1071.

[13] J. Anandan, New Relativistic Gravitational Effects Using Charged-Particle Interferonometry, *General Relativity and Gravitation,* 16(1);1984, 33-41.

[14] Torr, D.G, N.Li, N., Gravitoelectric-electric coupling via superconductivity, *Foundations of Physics Letters*, . 6(4); 1993, 371-383.

[15] Musha,T., Sawatari, K., Possibility of space drive propulsion by high potential field, Proc. of the 36th *Space Sci. and Tech. Conf.*, JSASS,1992, 95-96(J).

[16] Musha, T., Abe,I., Biefeld-Brown effect and electro-gravitic propulsion by high potential electric field, *Proc. of the 24th JSASS Annual Meeting, JSASS*; 1993, 189-192(J) .

[17] Brillouin, L., Relativity Reexamined, Academic Press. Inc, New York, 1970.

[18] Musha, T., Explanation of dynamical Biefeld-Brown Effect from the standpoint of ZPF field, Proceedings of the Fifth IAA Symposium on Realistic Near-Term Advanced Scientific Space Missions, Aosta, Italy, Jul.2-4, 2007.

[19] Musha, T., Explanation of Dynamical Biefeld-Brown Effect From The Standpoint of ZPF Field, *JBIS*, Vol.61; 2008, 379-384.

[20] Lerner, E.J., The Big Bang Never Happened, Vintage Books, New York, 1991.

[21] Thornhill, W., Talbot, D., The Electric Universe, Mikamar Publishing, Portland, OR, 2007.

[22] V.A.Bailey, V.A., Existence of Net Electrical Charges on Stars, *Nature,* 186; 1960, 508-510.

[23] Jefimenko, O.J., *Causality Electromagnetic Induction and Gravitation*, Electric Scientific Company, Star City, West Virginia, 1992.

*(J) means the report is written in Japanese.

Chapter 3

DYNAMICAL BIEFELD-BROWN EFFECT FROM THE STANDPOINT OF ZPF FIELD

Takaaki Musha*

Advanced Sci.-Tech. Research Organization,
Yokohama, Japan

ABSTRACT

The research group of the HONDA R&D Institute observed the weight reduction at an experiment by applying an alternate electric field to a capacitor. This phenomenon, which can be called as "dynamical Biefeld-Brown effect", cannot be explained within the framework of conventional physics. From the standpoint of the ZPF field, the author tries to explain this phenomenon as an interaction between the vacuum electromagnetic zero-point field and the high potential electric field. By the theoretical analysis, it is considered that the interaction of zero-point vacuum fluctuations with a high potential electric field can induce a greater momentum for the dielectric material, which would produce a sufficient artificial gravity to permit interstellar travel for a potential spacecraft.

* E-mail address: takaaki.mushya@gmail.com.

1. INTRODUCTION

In 1956, T.T.Brown presented a discovery known as the Biefeld-Bown effect (abbreviated B-B effect) that used a sufficiently charged capacitor with dielectrics that exhibited unidirectional thrust in the direction of a positive plate [1].

In recent years, it was confirmed by the research group of the HONDA R&D Institute that the weight reduction was observed by applying a high intensity electric field to the capacitor [2]. D. R. Bueler performed these experiments with high voltage single and parallel plate electric capacitors, which indicated a net force acting upon the capacitor mass [3]. The transient mass shift was also observed with the experiment conducted by J.F.Woodward for the dielectric material by rapidly charging high energy capacitors [4].

Recently it has been reported that Borbás Miklós detected a force by using a device, which was a simple ping-pong ball with a high voltage discharge element inside the ball [5]. From his experimental result, the device generated a propulsive force which was not due to air movement caused by ion flow (ion-wind) contrary to the conventional explanation that the generated force is due to the ion transfer between the plates of the capacitor. M. B. King gave an idea in his book [6] that electromagnetic fluctuations of the vacuum was the source of the B-B effect, which generates the unidirectional thrust for the dielectric material under a high potential electric field.

He considered that a slight coherence of vacuum fluctuations due to the high potential electric field caused an alternation of inertial properties of the body with the ionic lattice of a rapidly spinning atom, but its mechanism was not fully explained.

H. E. Puthoff proposed an idea in his article [7] that gravity is a form of long-range van der Waals force associated with the rapid motion of elementary particles (Zitterbewegung) in response to zero-point fluctuations (ZPF) of the vacuum. He suggested that if one could somehow modify the vacuum medium, then the mass of a particle or object in it would change according to the zero-point field theory [8].

By using the theory of the ZPF field, the author tries to show that the impulsive electric field applied to the dielectric material may produce a sufficient artificial gravity to permit interstellar travel for the potential spacecraft used in this chapter.

2. Experiments for Electro-Gravity Effect

1) Borbás Miklós Experiment in a High Vacuum

Borbás Miklós has detected a force that appears counter to the typical 'lifter' theory. A lifter is an electrohydrodynamic device utilizing an electrical phenomenon known as the Biefeld–Brown effect to produce thrust in the air, without requiring any combustion or moving parts. In its basic form, it simply consists of two parallel conductive electrodes, one in the form of a fine wire and another which may be formed of either a wire grid, tubes or foil skirts with a smooth round surface. When such an arrangement is powered up by high voltage in the range of a few kilovolts, it produces thrust.

His device is a simple ping-pong ball with a high voltage discharge element inside. He said that the device generated a propulsive force which is not due to air movement caused by ion flow (ion-wind).

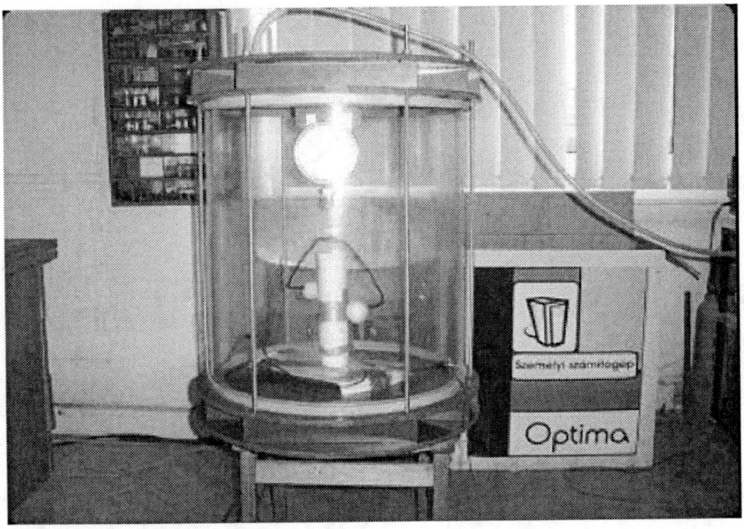

Figure 1. Borbás Miklós Experiment.

Miklòs believes the thrust effect is caused by the exchange of energy from the Zero Point field which is the foundation of reality and all energy fields. He has a significant deep thought on how the ZPE interacts with ions and atomic nuclei, and the story winds back to using Mach's Principle at play in the sub-atomic level... actually in the electron valence shells where quantum electrodynamics exchanges energy into and out of the aether, and into the nucleus that gives up its hold on the ZPE that is the principle cause of a change in inertia.

2) NASA's Experiment in a Vacuum Chamber

Here are some comments about the asymmetrical capacitors tested in a High Vacuum:

These tests were performed in the NASA NSSTC LEEIF vacuum chamber at the Marshall Space Flight Center in Huntsville (Alabama). The vacuum is at a High Vacuum at Torr, this is equal to the vacuum space conditions at about 350 km far from the earth's surface. For information, flying at an altitude of 100 km is equal to space flight... In the video of the tests, two asymmetrical capacitors are mounted on a rotating arm with a torsion wire used as a rotational axis. A potential difference between the wire and the main armature of the asymmetrical capacitor is slowly increased from 0 to +45 KV. These tests have been conducted by Gravitec Inc. and NASA on July 3, 2003. These results have not been published until today due to confidential reasons.

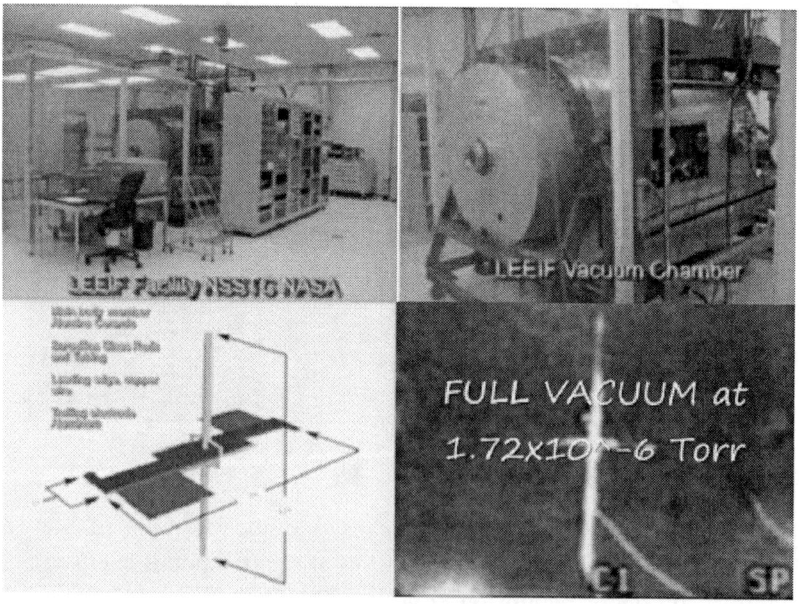

Figure 2. NASA Experiment in a Vacuum Chamber [9].

Tests results are shown as follows:

At the atmospherical pressure, it was observed that a thrust is in the plane of rotation and directly applied on the asymmetrical capacitors when the voltage is increased from 0 to +45 KV. This produces a torque on the apparatus. When the voltage is back down to 0 voltage, the device retrieves its initial position.

At the pressure of 1.72×10^{-6} Torr (High Vacuum conditions), the apparatus rotates when the High Voltage is increased from 0 to +45 KV. However, the observed thrust is weaker than at the atmospheric pressure. When the voltage is back down to zero voltage, the device retrieves its initial position. The oscillation of the apparatus in the horizontal plane is amplified with a parametrical effect due to the superposition of these impulses.

3) JNL Experiment in the Oil

The purpose of this experiment is to test the Townsend Brown Electrokinetic apparatus in different mediums such as mineral oil.

This experiment is fully described in the US Patent No 2949550 filed on Aug 16, 1960 and titled: "Elektrokinetic Apparatus". This device uses the same working principle as the Lifter devices: It is therefore an object of my invention to provide an apparatus for converting the energy of an electrical potential directly into producing a mechanical force suitable for causing relative motion between a structure and the surrounding medium.

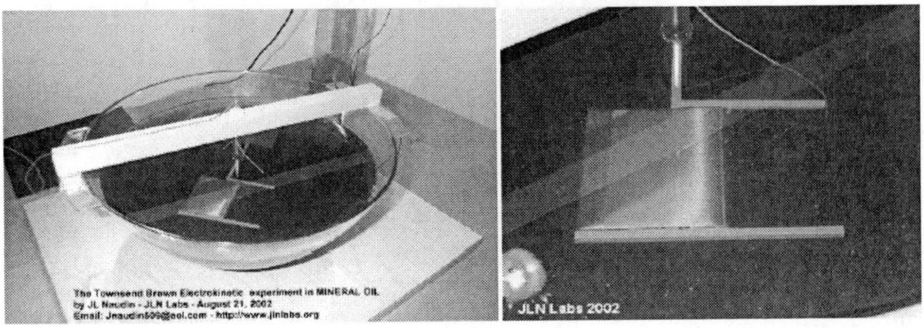

Figure 3. JNL Experiment in the Oil [10].

This device uses two rectangular asymmetrical capacitors made with aluminum. They are wing shaped (60 x 50 mm) and a thin enameled wire (0.05 mm diam) is placed at 35 mm from the leading edge. These small wings are mounted on a frictionless rotor (70 mm diameter). The rotating T. T.

Brown wings are placed in a plexiglass vessel which can be filled with a liquid.

When the HV power supply was switched on (37.7 KV DC, 810 uA), this devised wings were set in motion. In spite of the high viscosity of the mineral oil used at a temp of 25°C, the rotor turns at a low speed (about 2 RPM).

4) HONDA Experiment by Applying Electric Field to the Capacitor

From the 1st of Feb. till the 1st of March in 1996, the research group of the HONDA R&D Institute conducted experiments to verify the B-B effect with an improved experimental device in 1996. This device would reject the influence of the corona discharges and electric wind around the capacitor by setting the capacitor in the insulator oil contained within a metallic vessel as shown in Figure 4 [11]. The capacitor used at the experiment used a circular plate made of a high permittivity dielectric glass with the thickness, $t = 1mm$ the diameter, and the weight, $W = 62\,g$.

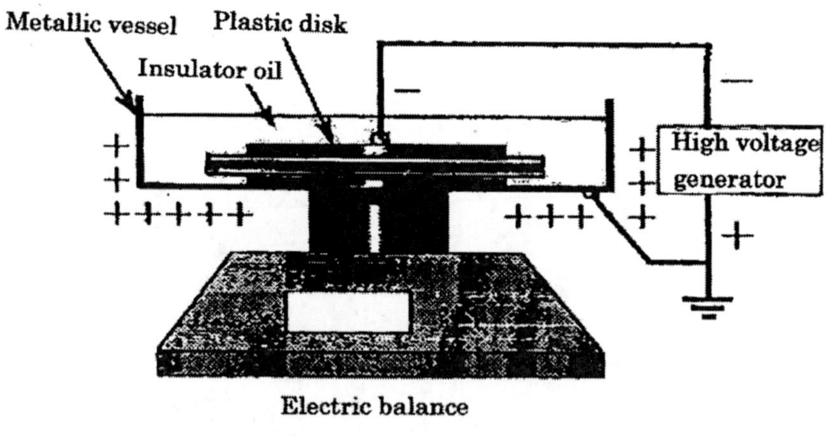

Figure 4. Schematic diagram of the experiment.

The electric balance used for the experiment had the resolution of 1mg. They conducted experiments for two cases, DC-18kV and AC-8kV pulses that were used to impress upon the capacitor by the experimental set-up as shown in Figure 5.

The circuit diagram for generating AC pulses is shown in Figure 6. The high AC voltage was supplied to the capacitor through an ignition coil to produce rectangular pulses with a repetition of 50 Hz. After amplifying AC pulses by the ignition coil, the impulsive electric field was applied to the capacitor through a diode by changing the polarity to produce minus-biased or plus-biased voltage to the capacitor, as shown in this figure.

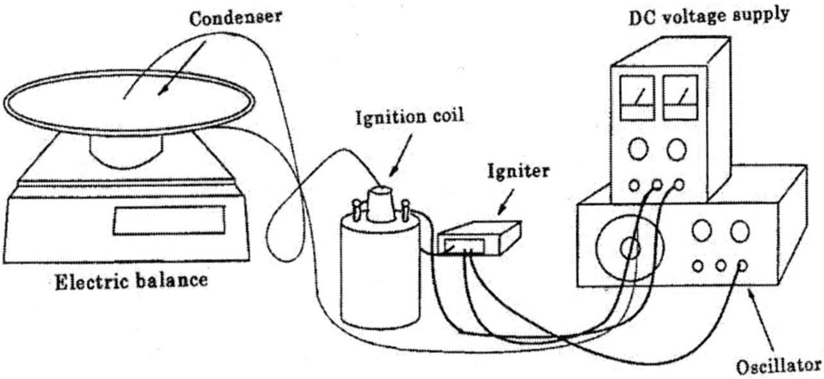

Figure 5. Experimental set-up for the experiment.

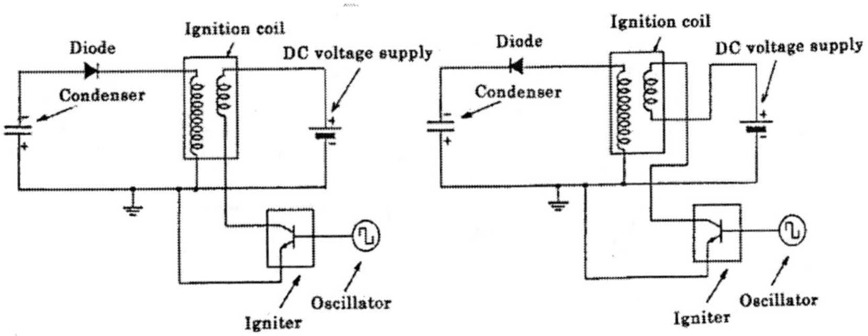

Figure 6. Circuit diagram for generating AC pulses.

The experimental results measured by the HONDA research group are shown in Figure 7. In this figure, DC (1) and DC (2) are for cases of DC-18kV, and PULSE (1) and PULSE (2) that are also used for cases of AC-8kV pulses, where the horizontal line is for the date of the experiment and the vertical line is for the change of the capacitor's weight.

From experimental results, it was found that the case of AC pulses exhibited higher reductions of weight by a factor of 2.77 than the case of DC exposures. The ratio of them becomes $\Delta W'/\Delta W = 2.77 \pm 0.87$, where is a weight reduction for AC pulses and is for the DC exposures. Maximum weight reduction was measured at the experiment by applying AC pulses with a weight reduction of 1.92 g, which was about 3% of the initial weight of the capacitor that was used at the experiment [2].

Supposing that the weight reduction of the capacitor is responsible for forces generated by the ionic transfer of the momentum, it can be estimated from the equations given by [11]:

$$\Delta M = \sqrt{\frac{2m_0 V_0}{q}}\frac{i}{g_0}, \qquad (1)$$

where is the mass of the carrier of the charge in the medium, is the charge of the carrier, is gravity acceleration on the Earth, is the applied voltage, and is the current density given by (C: capacity, f: frequency of applied voltage).

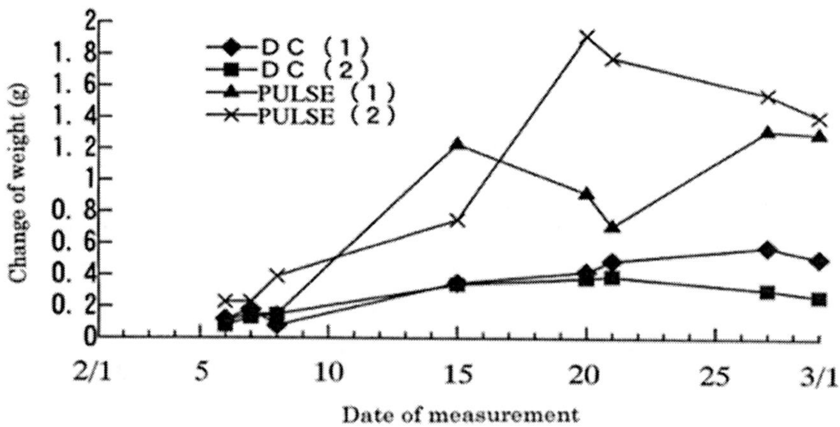

Figure 7. Measured results at the experiment.

By introducing experimental values for AC cases into Eq. (1), where $C = 1682$ pF, $V_0 = 8\,\text{kV}$ and $f = 50$ Hz, the mass reduction due to ionic winds becomes $\Delta M = 1.3 \times 10^{-3}$ g, which is negligibly small compared with the experimental result.

The author obtained the following formula for the electrogravitic effects induced on a dielectric material from a weak field approximation of Einstein's general relativity theory [12].

$$E_g \approx -Z\sqrt{4\pi\varepsilon\, G}\cdot E = -8.62\times 10^{-11} Z\sqrt{\varepsilon_r}\cdot E\ , \tag{2}$$

where is a magnitude of electric field impressed upon the capacitor, is a number of electrons circulating around the atomic nucleus, is a permittivity of the dielectric material, is a specific inductive capacity of the dielectric material, and is the gravitational constant. B. Ivanov of the Institute for nuclear Research and Nuclear Energy in Bulgaria also found the similar equation deduced from Weyl Majumdar-Papapetrou solutions of the general relativity theory, which is equivalent (modulo Z) to the author's equation, as shown in the Appendix.

From which the weight reduction of the capacitor under impressed electric field given by can be estimated by:

$$\Delta W = -M\cdot E_g = 8.62\times 10^{-11} Z\sqrt{\varepsilon_r}\, EM\ , \tag{3}$$

where is a mass of the capacitor.

By introducing values of the experiment, $Z = 10$, $\varepsilon_r = 10$, $M = 62$ g, into Eq.(3), we obtain $\Delta W = 0.31$ g for DC-18kV exposures, which is close to $\Delta W = 0.29 \pm 0.17$ g obtained from the experiment, but we have $\Delta W = 0.14$ g for AC-8kV pulses, which is much lower than the experimental result, $\Delta W' = 0.88 \pm 0.63$ g.

Thus it is considered that the B-B effect for AC-8kV pulses might be due to another mechanism other than the ion transfer of momentum or a weight reduction generated by the electrogravitic effect.

3. Mass Shift Created by the External Electromagnetic Field

According to the quantum electrodynamics, the quantum vacuum is filled with the electromagnetic fluctuations of the zero-point field (ZPF). Haish, Rueda and Puthoff pointed out in their paper [13] that the formula on the ZPF-determined inertial mass is remarkable similarity to the mass shift derived from the quantum theory. Under the intense electromagnetic field, it has been predicted that the electron experiences an increase in its rest mass under an intense electromagnetic field.

The analogical formula with respect to the mass shift of the electron under intense electromagnetic field was discovered by P. Milonni shown as follows [14]: Let be the electrodynamic Hamiltonian of the particle under high electromagnetic field, it has the form shown as:

$$H_A = \frac{e^2}{2m_0 c^2} < A^2 > . \tag{4}$$

The similar equation by using terms of the ZPF field was also proposed by Haish, Rueda and Puthoff shown as [13]:

$$H'_A = \frac{e^2 \hbar}{2\pi m_0 c^3} \omega_c^2 , \tag{5}$$

where is a Plank constant divided by 2π and is a cutoff frequency of ZPF spectrum in the vacuum.

Assuming that electrodynamic Hamiltonians shown in Eqs. (4) and (5) are identical with each other, then we have for the dielectric material under impressed electric field as shown in Figure 8. We suppose that the cutoff frequency of the vacuum is shifted as when the electromagnetic field is impressed on the dielectric material, becomes:

$$\Delta H'_A = \frac{e^2 \hbar}{2\pi m_0 c^3} \{(\omega_0 + \Delta\omega)^2 - \omega_0^2\} \approx \frac{e^2 \hbar}{\pi m_0 c^3} \omega_0 \Delta\omega , \tag{6}$$

where is the Plank frequency given by $\omega_0 = \sqrt{c^5 / \hbar G} \approx 3 \times 10^{43}$ Hz.

As shown in this figure, at the initial state, then we obtain the formula given by:

$$\Delta\omega \approx \frac{\pi\ c}{2\hbar\omega_0} < A^2 > . \tag{7}$$

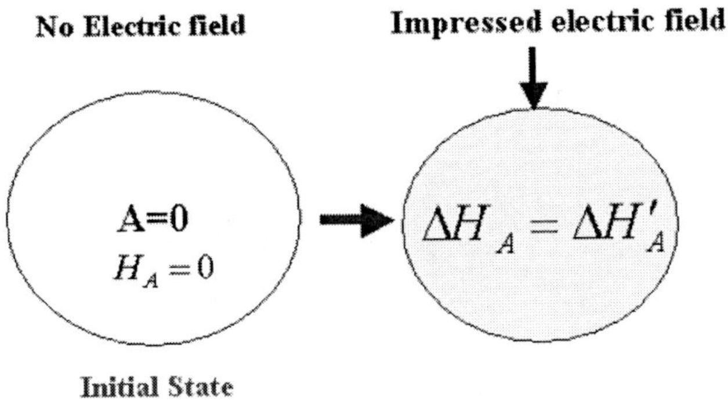

Figure 8. Electrodynamic Hamiltonian under impressed electric field.

According to the gravitational theory proposed by Haish, Rueda and Puthoff , we can suppose that the inertial mass of elementary particles induced by the ZPF field can be given by

$$m = \frac{\Gamma \hbar \omega_c^2}{2\pi\ c^2}, \tag{8}$$

where is the radiation reaction damping constant defining the interaction of charged elementary particles with electromagnetic radiation field.

From which, we have:

$$\Delta m / m = \frac{2\Delta\omega_c}{\omega_c} = \frac{\pi\ c}{\hbar\omega_0^2} < A^2 > . \tag{9}$$

For the dipole field generated by the variance of electric charge as shown in Figure 9, the vector potential of the electromagnetic field becomes [15]:

$$A = \frac{1}{4\pi\varepsilon_0 c^2} \frac{\dot{p}(t - r/c)}{r} \approx \frac{1}{4\pi\varepsilon_0 c^2} \frac{\dot{p}(t)}{r}, \tag{10}$$

where is a dipole momentum given by $p = qd$ (q : charge of particles, d : displacement of the charge).

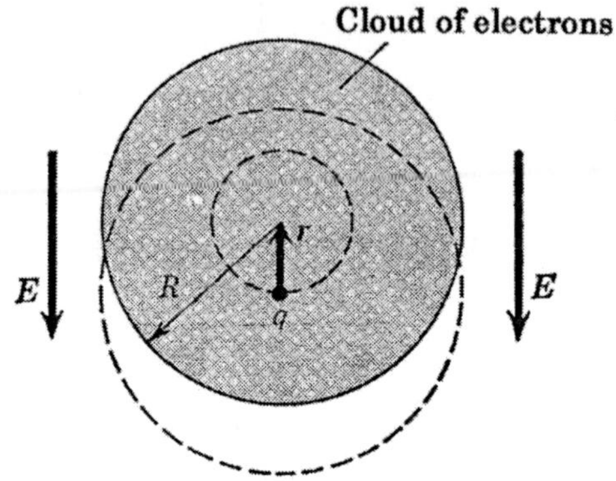

Figure 9. Dipole field generated by the alternating electric field.

If we let $p(t) = p_0 \sin \omega t$ ($p_0 = Ned$), then we have the vector potential as:

$$A = \frac{1}{4\pi\varepsilon_0 c^2} \frac{\omega\, p_0 \cos \omega t}{r} = \frac{1}{4\pi\varepsilon_0 c^2} \frac{\omega Ned \cos \omega t}{r}, \tag{11}$$

In this equation, is a number of charges per unit volume and is given by

$$d = \frac{e}{m} \frac{E}{\omega_e^2 - \omega^2}, \tag{12}$$

where is an alternating electric field and is a resonant angular frequency given by (α_e: electron polarizability), which yields about $10^{15} \sim 10^{16}$ Hz. As the energy dissipation can be incorporated into the analysis by replacing real quantities with complex ones as shown with $\omega \to \omega(1 + i\eta / 2)$, where is a damping factor, we obtain the following equation from above equations for the charged sphere with a radius of R:

$$\Delta M(\omega)/M = \frac{\pi G}{c^4} \int < A^2 > dv$$

$$= \frac{1}{16\pi} \frac{N^2 e^4 G}{\varepsilon_0^2 c^4 m^2} \frac{\omega^2}{(\omega_e^2 - \omega^2)^2} E^2 \int_0^{\pi} \sin \theta d\theta \int_0^R dr \int_0^{2\pi} d\varphi$$

$$= \frac{N^2 e^4 G}{4\varepsilon_0^2 m^2 c^8} \frac{\omega^2}{(\omega_e^2 - \omega^2)^2 + \eta^2 \omega^4} E^2 R \quad . \tag{13}$$

For the impulsive electric field, which has a wide frequency range of spectrum, ($\omega_2 - \omega_1$) is large compared to the width of the resonance, the following integration over frequencies across the resonance becomes:

$$\int_{\omega_1}^{\omega_2} \frac{\omega^2 d\omega}{(\omega_e^2 - \omega^2)^2 + \eta^2 \omega^4} = \frac{\pi}{2\eta\omega_e} \quad . \tag{14}$$

From which, we obtain the ratio of the mass shift vs. its rest mass under the impulsive electromagnetic field given by:

$$\Delta M / M = \int_{\omega_1}^{\omega_2} \Delta M(\omega) / M \cdot d\omega = \frac{\pi}{8} \frac{e^4 G}{\varepsilon_0^2 m^2 c^8} \frac{N^2 R}{\eta \omega_e} E^2 \cdot \tag{15}$$

Assuming that $\eta = 2e^2 / 3mc^2$, that equals the Abraham-Lorenz damping constant [16], we have:

$$\Delta M / M = \frac{3\pi}{16} \frac{e^2 G}{\varepsilon_0^2 m c^6} \frac{N^2 R}{\omega_e} E^2 , \tag{16}$$

where is a number of electrons per unit volume in a space including the dielectric material and is a radius of the electron cloud and is a magnitude of the impulsive electric field.

4. The Possibility to Produce Artificial Gravity by an Electromagnetic Field

Combining Eqs. (2) and (16), the force produced by the electrogravitic field becomes:

$$F = -(M + \Delta M)E_g \approx Z\sqrt{4\pi\varepsilon G}\left(1 + \frac{3\pi}{16}\frac{e^2 G}{\varepsilon_0^2 mc^6}\frac{N^2 R}{\omega_e}E^2\right)M \, , \quad (17)$$

where is a gravitational field induced by the external electric field.

From this equation, it is seen that the electrogravitic effect generated for the dielectric material can be amplified by the impulsive electric field as observed at the Honda experiment. This also supports the Brown's experiment where he observed a large thrust that was associated with a spark between electrodes over the disc structure that also consists of the dielectric material and electrodes as shown in Figure 10.

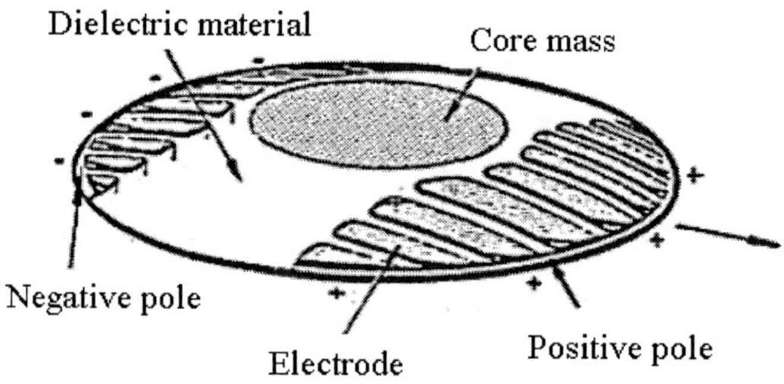

Figure 10. Structure of the electrogravitic craft by T. T. Brown.

J. L. Naudin also proposed the electrogravitic craft that was named, "The ARDA Morning Star" that included electro-aerodynamic surfaces with a HV leading edge in 2000 shown in Figure 11.

From Eq. (17), the new factors that increase the force for the dielectric material are presented to:

- Increase the magnitude of an electric field impressed on the dielectric material.
- Increase the number of electrons per unit volume in a space that includes the dielectric material.
- Increase the radius of the electron cloud in a space that includes the dielectric material.

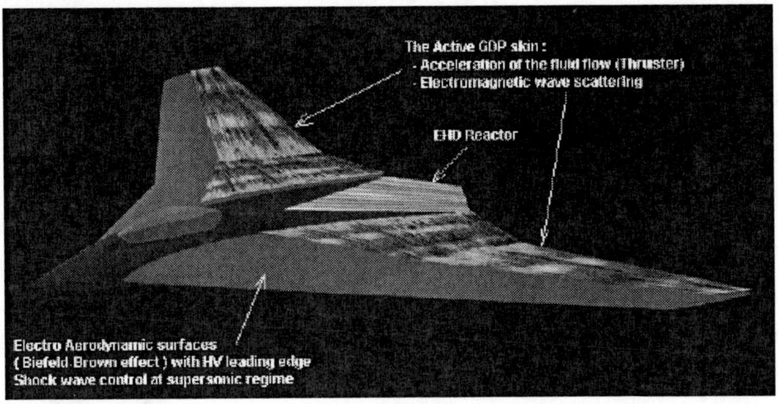

Figure 11. Electrogravitic craft proposed by J. N. L. Naudin (http://jnaudin.free.fr/html/morningstar.htm).

As the mass shift of the dielectric material under the electrogravitic field satisfies $\Delta W = -M \cdot E_g$, then we have:

$$\frac{\Delta W'}{\Delta W} \approx \frac{E'_g}{E_g} = \left(1 + \frac{3\pi}{16}\frac{e^2 G}{\varepsilon_0^2 mc^6}\frac{N^2 R}{\omega_e}E'^2\right)\frac{E'}{E}, \tag{18}$$

where and are weight shifts observed for AC pulses and DC exposures, and and are amplitudes of AC or DC electric field impressed respectively on the capacitor.

By introducing experimental values, $E = 18 \times 10^6$ V/m $E' = 8 \times 10^6$ V/m and $R = 0.5$ mm into Eq.(18), where we let $\omega_e = 10^{15}$ rad/s and $m = 9.11 \times 10^{-31}$ kg (for the electron's mass), we obtain $\Delta W'/\Delta W = 3.2$, when we suppose that $N = 10^{26}$, which is lower than the critical plasma density that is at $N = 10^{28}$.

This is close to the experimental result obtained by the Honda research group and hence it can be considered that the electric discharge between plates, which has a wide range of spectrum, produces a great force under the high impulsive electric field. As the mass shift predicted by Eq. (16) is dependent on the external electric field, the variance of weight reduction is observed at the experiment that might be due to the secondary electrostatic effect related to the global electric field as reported by G. V. Stephenson [17], the influence of which cannot be totally rejected by the electromagnetic shielding.

From the equation for the momentum given by $f = dP / dt$, the momentum produced by the electrogravitic field becomes:

$$P_{field} = \int F dt = \int m \cdot E_g dt \approx m \cdot E_g \Delta t \,. \qquad (19)$$

When we let $\Delta t = l / v_d$, where is a separation between electrodes and is a drift velocity of electrons, the momentum produced by the impulsive electric field becomes [18]:

$$P_{field} = \int (M + \Delta M) \cdot E_g dt \approx Z\sqrt{4\pi\varepsilon_0\varepsilon_r G}M\left(1 + \frac{3\pi}{16}\frac{e^2 G}{\varepsilon_0^2 mc^6}\frac{N^2 R}{\omega_e}E^2\right)\frac{El}{v_d} \,. \qquad (20)$$

Eq. (20) suggests that high coronal discharge around the spacecraft which consists of a dielectric material with a highly charged density of electrons that would induce a gravitational force potentially for space propulsion. According to this equation, the greater the radius of the electron cloud, the greater the momentum will be produced, hence the space propulsion system for the spacecraft as shown in Figure 12 can be proposed [19].

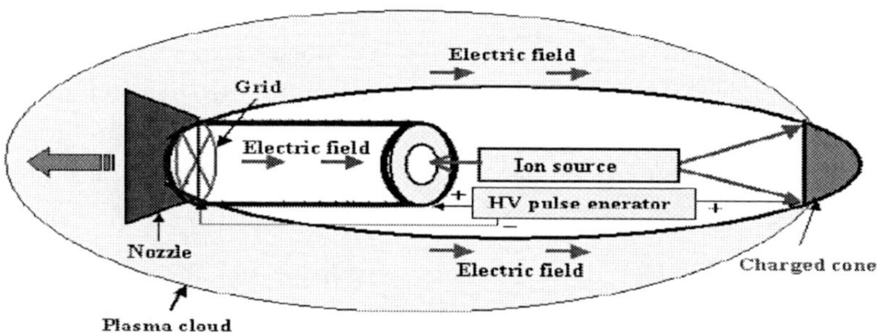

Figure 12. Space propulsion system by utilizing the ZPF field.

If we suppose that the electromagnetic field is generated by a plasma cloud that surrounds the spacecraft that can influence its mass according to Eq. (20), it can attain a higher capability of acceleration by using coronal discharges that are generated by an on-board high-voltage (HV) pulse generator as shown in Figure 12.

This spacecraft is powered by the HV pulse generator to produce negative ion clouds around the body. The high-voltage power is applied to the cone of the craft and the exhaust nozzle to produce a greater potential difference between the cone and the exhaust nozzle. Hence this ZPF propulsion system generates a high intensity electric field along its body to produce a locally altered gravity field, which is coupled with the mass shift generated by the coronal discharge that causes a greater forward directed gravitic-force acting upon the spacecraft. Figure. 13 is for the ZPF propulsion mode is surrounded by a negative ion cloud.

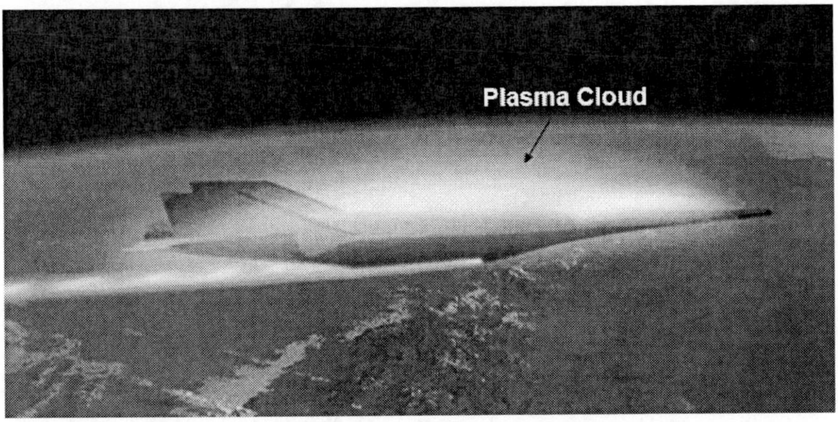

Figure 13. Artist's view of the ZPF-Plasma propulsion system.

From Eq. (20), the velocity of the spacecraft generated by the high intensity electric field can be estimated from $v = P_{field} / M'$ (M': total mass of the spacecraft) [19].

Assuming that the electro-dynamical damping factor has the value on the order of the Abraham-Lorenz damping constant, it can be seen from the calculation results by assuming that $l = 10$ m, $R = 5$ m, $\omega_e = 10^{15}$ rad/s, $m = 9.11 \times 10^{-31}$ kg (electron's mass), and $N = 10^{26}$, $v_d = 10^8$ m/s for the value of the vacuum arc [20]. Assuming $M \approx M'$, this spacecraft has the possibility to attain the velocity of 9.5×10^5 km/h as shown in Figure 14, when applied to 10GV impulsive voltage to the electrodes, which is over 30 times the speed generated by a conventional chemical rocket, which enables us to reach the stars far beyond the rim of our solar system.

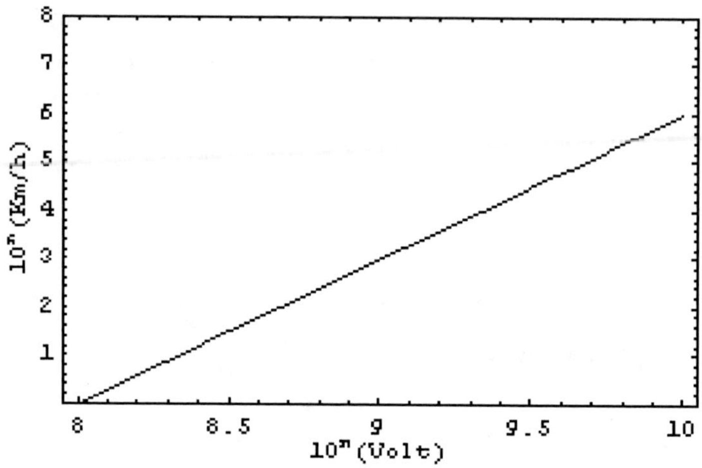

Figure 14. Velocity produced by the impulsive electric field.

Hence it is considered that this electrogravitic system may permit interstellar travel for the spacecraft instead of using chemical rockets when the ether of the following factors is increased;

1. voltage is impressed to the dielectric material,
2. separation occurs between electrodes,
3. the number of electrons per unit volume in space including the dielectric material, and
4. the radius of the electron cloud.

Frank Wilczek, Nobel lautriate in physics, also proposed an idea similar to Puthoff that space consists of the "Grid" which is made up of spontaneous activity that occurs between magnetic and electric fields [21]. He claimed that space grids can produce an exothermic reaction, releasing more energy than is initially used. Thus space is filled to the point that we can utilize a propulsion concept for a feasible space propulsion system.

CONCLUSION

From the theoretical analysis using the zero-point field theory, it is considered that the origin of the dynamical Biefeld-Brown effect might be attributed to the interaction between the zero-point vacuum fluctuations and a high potential electric field that is impressed upon a capacitor. The theoretical analysis result suggests that the impulsive electric field applied to the dielectrics, which consists of the spacecraft would produce an artificial gravity sufficient for practical application to allow interstellar travel.

REFERENCES

[1] INTEL, *Toward Flight without Stress or Strain or Weight*, Interavia, XI; 1956, 373-374.

[2] Musha, T., Theoretical Explanation of the Biefeld-Brown Effect, *Electric Spacecraft Journal*, Issue.31; 2000, 29.

[3] Buehler, Exploratory Research on the Phenomenon of the Movement of High Voltage Capacitors, *Journal of Space Mixing*, 2; 2004, 1-22.

[4] Mahood, T.L., Propellantless Propulsion: Recent Experimental Results Exploiting Transient Mass Modification, *Space Technology and Applications International Forum*, 1999, 1014-1020.

[5] Milkos, B., www.fw.hu/bmiklos 2000/unipolar.html.

[6] King,M.B., Is Artificial Gravity Possible?, Tapping The Zero-Point Energy, Adventures Unlimited Press, Kempton, Illinois, 2002.

[7] Puthoff, H.E, Gravity as a zero-point-fluctuation force, *Physical Review A,* Vol.39, No.5; 1989, 2333-2342.

[8] Puthoff, H.E., Can the vacuum be engineered for spaceflight applications? Overview of theory and experiments, *Journal of Scientific Exploration,* Vol.12, No.1; 295-302, 1998.

[9] Naudin, JL, NASA *Experiment in a Vacuum Chamber*, jnaudin.free.fr/ lifters/ascvacuum/index.htm.

[10] Naudin, JL., *The Townsend Brown Electrokinetic experiment in Mineral Oil,* jnaudin.free.fr/html/advprop.htm, 2002.

[11] Bahder, T.B., Fazi, C., Force on an Asymmetric Capacitor, Army Research Laboratori, March, 2003, http//arXiv.org/ftp/physics/ papers/0211/0211001.pdf.

[12] Musha, T., Explanation of Dynamical Biefeld-Brown Effect from the standpoint of ZPF field, Proceedings of the fifth IAA Symposium on Realistic Near-Term Advanced Scientific Space Missions, Aosta, Italy, 2007.

[13] Haisch,B., A.Rueda, A. and Putthoff, H.E., *Advances in the Proposed Electromagnetic Zero-Point Field Theory of Inertia*, arViv:physics /9807023 v2 (22.Jul.19998)

[14] Haish,B., Rueda, R. and Puthoff,H.E, Inertia as a zero-point-field Lorenz force, *Physical Reviw. A* , Vol.49,No.2; 1994, 678-694.
[15] Feynmann, R.P., .Leighton, R.B. and Sands,M., *The Feynman Lectures on Physics,* Vol.II, Addison-Wesley, New York, 1964.
[16] Jackson, J.D., *Classical Electrodynamics*, Chapter.17, John Wiley & Sons, Inc., New York, 1975.
[17] Stephenson, G.V., The Biefeld Brown Effect and the Global Electric Circuit, *Space Technology and Applications International Forum*; 2005, 1249-1255.
[18] Musha, T., Explanation of Dynamical Biefeld-Brown Effect from the Standpoint of ZPF Field, *JBIS*, Vol.61; 2008, 379-384.
[19] Musha, T., Possibility of the Space Propulsion System Utilizing the ZPF Field, *Proceedings of Space, Propulsion & Energy Sciences International Forum;* 2009, 194-201.
[20] Boxman, R.L., Martin, P.J., Sanders, D.M., *Handbook of Vacuum Arc Science and Technology: Fundamental and Applications*, Noyes Publications, Park Ridge, N.J.1996.
[21] F.Wilczek, The Lightness of Being: Mass, Ether, and the Unification of Forces, *Basic Books*, New York, 2010.

Chapter 4

OTHER ELECTROMAGNETIC GRAVITY CONTROL DEVICES

Takaaki Musha**[*]**

Advanced Sci.-Tech. Research Organization,
Yokohama, Japan

ABSTRACT

There are realms that stretch the boundaries of known relativity and quantum theories. Some of these attempts incorporate electromagnetism or the relativity theory, while others attempt to rewrite existing theories. Some of these interesting theories which can be utilized for space travel are presented in this chapter.

1. INTRODUCTION

The Russian researcher Eugene Podkletnov claims to have discovered, whilst experimenting with superconductors in 1995, that a fast rotating superconductor reduces the gravitational effect. Many studies have attempted to reproduce Podkletnov's experiment, always to no results. In 1989, Ning Li, of the University of Alabama in Huntsville theoretically demonstrated how a time dependent magnetic field could cause the spins of the lattice ions in a

[*] E-mail address: takaaki.mushya@gmail.com.

superconductor to generate detectable gravitomagnetic and gravitoelectric fields. In 1999, Li and her team appeared in Popular Mechanics, claiming to have constructed a working prototype to generate what she described as "AC Gravity." No further evidence of this prototype has been offered.

Robert L. Forward, Hughes Research Laboratories, Malibu, described the theoretical generation of dipole gravitational fields by accelerating a super-dense fluid through pipes wound around a torus. Such techniques, though theoretically sound, have been far beyond the reach of technology.

Surveys conducted by engineers of physical principles as potential foundations for gravity control propulsion theories conferred high ratings for Burkhard Heim's work. Comments and investigations by physicists into his initial and extended propulsion theories appeared during 2008. Heim's earliest paper was completed in 1956 as a progress report that was archived at the Gravity Research Foundation. His theory was then published in a series of four papers during 1959 by a little known aerospace engineering journal. According to Helmut Goeckel's preface to the first paper in the series, unbridled interest of unsavory firms had caused Heim to cease from further developments of his propulsion theory. He indicated various aerospace and ordnance companies had made several attempts to kidnap him. Heim devoted the remainder of his life to refining the unified field attributes of his theory.

2. Gravitational Effect Induced by Coupling via Superconductivity

The Russian material scientist named Dr. Eugene Podkletnov claimed that he found an antigravity effect while working with a team of researchers at Tampere University of Technology in Finland. They made the devise and confirmed significant weight loss by using the ring of superconducting ceramics spinning at 5000 rpm, which is shown in Figure 1[1,2]. The rotating disk showed a variable loss of weight from less than 0.5% to 2%.

Following to these results, the experiment named the Delta-G experiments were carried by the Marshall Space Flight Center in USA which attempted to duplicate a claimed anomalous weight loss up to two percent in objects of various compositions suspended above a rotating 12-inch diameter Type II ceramic superconductor.

American scientists, Douglas Torr and Ning Li, at the University of Alabama proposed the theory to explain the antigravity effect as follows[3]: In the weak field low-velocity limit of general relativity, it is known that the field equation can be written approximately in the form of Maxwell's equations.

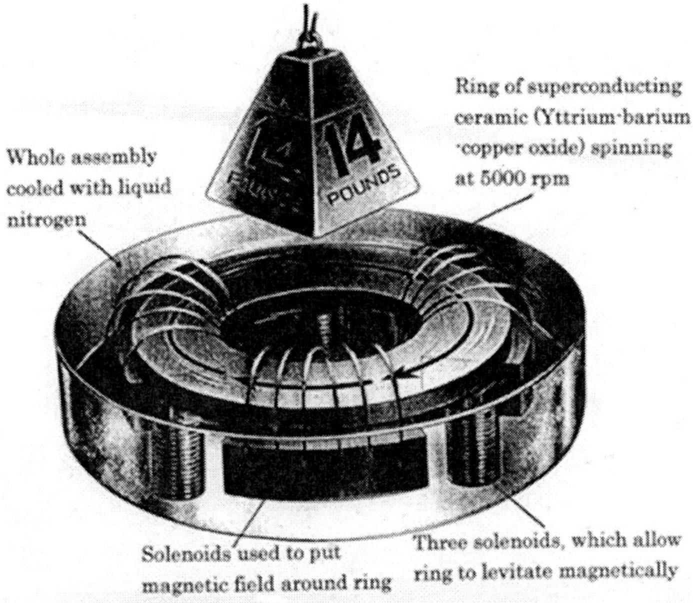

Figure 1. Device to generate gravitational field by using a mass suspended over a superconductor disc.

We let be the gravitoelectric field and be the gravitomagnetic field, those satisfy and $B_g = \nabla \times A_g$, where and are the gravito scalar potential and London gauge potentials.

Then the equation of motion of a Copper pair of mass with a charge moving with the velocity in electric and magnetic fields and can be approximated by a generalized Lorentz force as

$$F = q(E + v \times B) + m(E_g + v \times B_g). \tag{1}$$

From which, it is considered that electromagnetism should have gravitational counterparts analogically.

$$\nabla^2 E = \frac{1}{\lambda_L}\left[E + \frac{m}{q}E_g + \nabla(\phi + \frac{m}{q}\phi_g)\right] + \mu_0 \varepsilon \frac{\partial^2 E}{\partial t^2} + \frac{1}{\varepsilon}\nabla \rho_e, \tag{2}$$

$$\nabla^2 E_g = -\frac{1}{\lambda_L^2}\frac{\mu_g}{\mu}\frac{m}{q}\left[E + \frac{m}{q}E_g + \nabla(\phi + \frac{m}{q}\phi_g)\right] + \mu_{g0}\varepsilon_g\frac{\partial^2 E}{\partial t^2} - \frac{1}{\varepsilon_g}\nabla\rho_m, \tag{3}$$

where and $\varepsilon_{go} = 1/4\pi G$.

By setting and $\widetilde{E}_g = -\partial A_g / \partial t$, they obtained

$$\widetilde{E} + \frac{m}{q}\widetilde{E}_g = \left(\frac{\partial A_0}{\partial t} + \frac{m}{q}\frac{\partial A_{g0}}{\partial t}\right)\cdot e^{-z/\lambda}, \tag{4}$$

where $\lambda^2 = \lambda_L^2[1-(\mu_g/\mu)m^2/q^2]^{-1}$.

Inside the superconductor, when it is satisfied the condition $z >> \lambda$, Eq.4 can be written as

$$\widetilde{E} + \frac{m}{q}\widetilde{E}_g = 0, \tag{5}$$

which means that the internal volume of a superconductor is a region of force-free field.

When $\partial A_{g0} / \partial t = 0$, there will be an electrically induced gravitoelectric field given by

$$\widetilde{E}_g = \frac{\mu_g}{\mu}\frac{m}{q}\widetilde{E}_0, \tag{6}$$

On the other hand, when the external magnetic vector potential is ramped up, a time dependent gravitomagnetic vector potential is generated as

$$\widetilde{E}_g = -\frac{\partial A_g}{\partial t} = \frac{\mu_g}{\mu}\frac{m}{q}\frac{\partial A_0}{\partial t}. \tag{7}$$

Notice that Ning Li had predicted one percent weight change in her original calculations, from which, it is shown that the coherent alignment of lattice ion spins will generate a detectable gravitomagnetic effect, and in the presence of a time-dependent applied magnetic vector potential field, a detectable gravitoelectric field.

Experiments conducted for a variety of different test masses were tested by applying electromagnetic field and it was reported of a weight loss up to 5%, which shows that there exists a coupling between static superconductors and gravity [4,5].

G. Modanese also made the attempt to provide a theoretical explanation of Podkletnov's observations.[6,7] He argued the shielding effect and slight expulsive force at the border of the shielded zone could be explained in terms of induced changes in the local cosmological constant. Modanese described several effects in terms of responses to modifications to the local cosmological constant within the superconductor [8].

Several groups around the world tried to replicate Podkletnov's gravity shielding observations [9], but those groups were not able to overcome the extremely challenging technical problems of replicating all aspects of the original Podkletnov's experimental conditions according to R.C. Woods [10]. Podkeltnov and Modanese also reported the generation of a beam of gravity-like impulses [11].

They indicated a high voltage discharge device had been constructed that emitted a horizontal, collimated beam, with sharp borders, of short impulses of a repulsive force field that could penetrate different bodies without any noticeable loss of energy. Analyses of the measurements indicated the impulses briefly caused accelerations one thousand times the rate of gravity.

Hence this may lead to the development of gravitational propulsion system.

3. Elecromagnetic Field Propulsion System

In establishing the electromagnetic nature of light, Maxwell opposed Weber's "action at a distance" with his dynamical model of a vacuum with hidden matter in motion. His ideas were expanded by Pointing through the energy-flux theorem.

Minkowski found, as a purely mathematical consequence of Maxwell's equations, that the Lorentz force density could be expressed as the divergence of Maxwell's tensor in vacuo, decreased by the rate of change of Poynting's vector as [12]

$$\rho E + \mu_0 J \times H = \nabla \cdot T_{vac} - \varepsilon_0 \mu_0 \frac{\partial}{\partial t} E \times H , \tag{8}$$

which can be interpreted as a local reaction force acting on charges and currents when the vacuum surrounding them is loaded with electromagnetic momentum.

G.M.Graham and D.G.Lahoz conducted measurement of forces related to electromagnetic momentum by using the cylindrical capacitor with a vacuum gap inside [13,14]. They observed changes in angular momentum, which agree with the classical theory with the error of 20%. This implies that the vacuum is the seat of something in motion wherever static fields are set up with non-vanishing Poynting vector and permanent magnets and electrets can be used to build a flywheel of electromagnetic energy steadily flowing in circles in the vacuum gap of a capacitor.

The notion of EM propulsion by the electromagnetic momentum is dated back to Joseph Slepian (1891-1969), who proposed the EM drive based on fields. He proposed a concept for electromagnetic spaceship by utilizing the momentum in a space under electromagnetic field shown as [15,16]

$$F = \frac{\partial D}{\partial t} \times B - \varepsilon_0 \mu_0 \frac{\partial E}{\partial t} \times H \tag{9}$$

or

$$F = \frac{\partial D}{\partial t} \times B - \varepsilon_0 \frac{\partial E}{\partial t} \times B \ . \tag{10}$$

Figure.2 shows the apparatus composed of a parallel plate capacitor between two solenoids electrically wired in series, which was suggested as a driving mechanism for a spaceship. He supposed that this devise composed of inductive-capacitive system supplied with 10^8 Hz power, would produce sufficient thrust of magnitude 1.5×10^7 dynes for the case $H = 5000$ gauss, ampere and the distance between plates being $0.03\,m$, if properly phased [17].

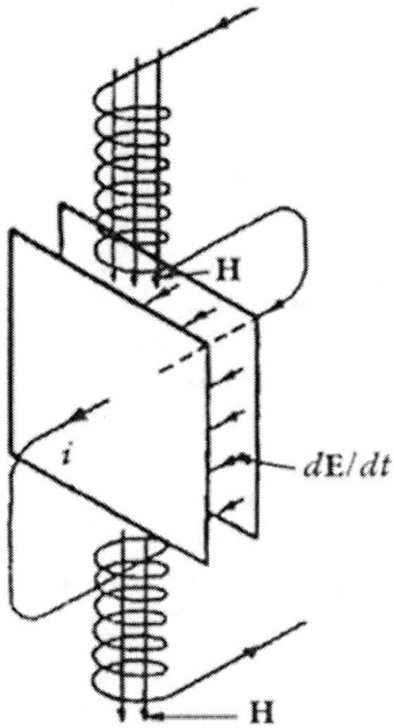

Figure 2. Schematic diagram of the Slepian's drive.

4. Heim's Theory on the Gravitophoton Propulsion

The Heim's quantum theory of gravity is based on the geometric view of Einstein, that the geometry itself is the cause of all physical interactions, but it uses the structure of Einstein's field equations only as a template for describing physical interactions in a higher dimensional discrete space, and extends them to the microcosm. Heim theory, initially proposed by the late German physicist Burkhard Heim, is an attempt to develop a theory of everything in theoretical physics. Heim attempted to resolve incompatibilities between quantum theory and general relativity. His theory requires extending spacetime with extra dimensions, an eight dimensional discrete space in which a smallest elemental surfaces exists [18, 19]. Within the quantum spacetime of Heim's theory, elementary particles are represented as "hermetry forms" or multidimensional structures of space. Heim has claimed that his theory yields particle masses directly from fundamental physical constants and that the resulting masses are in agreement with experiment. For Heim, this composite nature was an expression of internal, six-dimensional structure. After his death, others have continued with his multi-dimensional "quantum hyperspace" framework.

Although it purports to unify quantum mechanics and gravitation, the original Heim's theory cannot be considered a theory of everything because it does not incorporate all known experimental data. In particular, it gives predictions only for properties of individual particles, without making detailed predictions about how they interact. The theory also allows for particle states that don't exist in the Standard Model, including a neutral electron and two extra light neutrinos, and many other extra states. Presently, there is no known mechanism for the exclusion of these extra particles, nor an explanation for their non-observation.

Predictions claimed to have been derived from first principles that are experimentally testable are: 1) Predictions of the masses of neutrinos; 2) Predictions of new particles; 3) Predictions of excited states of existing particles, and 4) Predictions for the conversion of photons into the so-called "gravito-photons" resulting in a measurable force. Heim's theory predicts that it should be possible to produce an effective "gravity" force by converting photons into "gravito-photons," a particle which he predicts the existence of, but has not yet been directly observed.

In the 1950s, Heim had predicted what he termed a 'contrabary' effect whereby photons, under the influence of a strong magnetic field in a certain configuration, could be transformed into 'gravito-photons', which would provide an artificial gravity force.

As of late 2006, groups at Berkeley and elsewhere were attempting to reproduce this effect. By applying their 'gravito-photon' theory to bosons, Dröscher and Häuser were able to predict the size and direction of the effect.

Heim attempted to resolve incompatibilities between quantum theory and general relativity. To meet that goal, he developed a mathematical approach based on quantizing space-time itself, and proposed the "metron" as a (two-dimensional) quantum of (multidimensional) space.

The Heim's theory is further developed by Walter Dröscher and Jochem Häuser. According to their paper, the fifth interaction, termed as graviton force, would accelerate a material body without the need of propellant. Gravitophoton interaction is a gravitation-like force, that come into both attractive and repulsive form.

Gravitophoton particles are generated in pairs from vacuum itself by the effect of vacuum polarization (virtual electrons), under the presence of a very strong magnetic field. For gravitation, Heim theory assumes that a gravitational potential arises from the gradient of a field $\phi(r)$. Position dependent mass is the function m, and r is the radial distance from a quanta of a point mass.

A differential equation used to describe the basis is

$$\left(\frac{d\phi}{dr}\right)^2 + 32\frac{c^2}{3}F\left(\frac{d\phi}{dr} + F\phi\right) = 0, \tag{11}$$

$$F = \frac{1}{r}\frac{h^2 + \gamma m^3 r}{h^2 - \gamma m^3 r} \tag{12}$$

If this equation is nondimensionalized the characteristic length of the system is

$$r_c = \frac{h^2}{\gamma m^3}. \tag{13}$$

The characteristic length is the distance from a point mass for which the field is $\phi(r) = 0$. It is also the case that the field attains its absolute minimum. Hence, the gravitational force is identically zero at this distance.

The solution to the differential equation has the curve concave up. The gravitational potential that arises from this field can be positive, negative or zero.

From which, the Heim-Lorentz equation can be obtained as

$$F_{gp} = -\Lambda_p e \mu_0 v^T \times H, \tag{14}$$

which describes the gravitational interaction resulting from negative gravitophotons, where the index p in indicates that only photon and neutron absorption processes are considered, is the velocity of the rotating torus, and is the magnetic field generated by the current loop.

Gravitophoton propulsion takes place in two phases. In phase one, a spaceship is subject to acceleration in the 4-dimentional space-time. Acceleration is achieved by the absorption of negative gravitophotons through the photons and neutrons in the torus material. Covering large interplanetary distances, would require the transition into parallel space, which is phase two of the field propulsion, involving the repulsive quintessence particle. A transition into a parallel space leads to an increase in speed by a factor n, compared to our 4-dimentional space-time.

The value is obtained by

$$n = (g_{gp}^{+} / g_{g}) \cdot (G_{gp} / G), \quad (15)$$

where is the field strength of the gravitophoton field, is the gravitational field strength and is a coupling constant of gravitophotons.

In the rotating torus, the positive and negative gravitophoton fields are generated together, and, because of energy conservation, their strength are equal and can be directly calculated from Eq.(14)

The Heim-Lorenz force equation can produce a force component in the vertical direction to the rotating torus, as shown in Figure 3 [16].

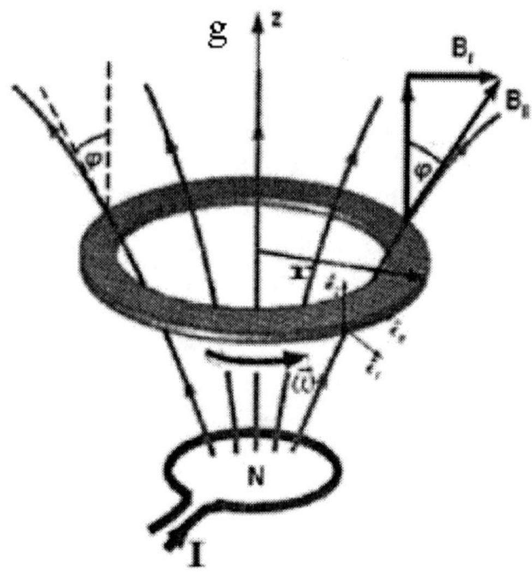

Figure 3. Physical principle of the devise to generate a gravitational field.

If his theoretical prediction is correct, the realization of a workable space propulsion device that can lift itself from the surface of the Earth can be realized. It has been estimated that the acceleration becomes larger than 1g for a spacecraft with the mass of kg, a magnetic induction of 20 T, a rotational speed of the torus of $v^{T} = 10^{3}$ m/s, and a torus mass of 2×10^{3} kg, which results in a total flight time to the Moon be 4 hours [19].

5. General Relativistic Gravity Machine Utilizing Electromagnetic Field

5.1. Forward's Gravitational Machine

There is an analogical relation between electromagnetic field and gravitational field in research papers by Harris and Braginsky[20,21].

A particle mass moving in a gravitational and electromagnetic field has the equation of motion as

$$\frac{d^2 x^\mu}{d\tau^2} + \Gamma^\mu_{\alpha\beta}\left(\frac{dx^\alpha}{d\tau}\right)\cdot\left(\frac{dx^\beta}{d\tau}\right) = \frac{e}{mc}F^\mu_\nu\left(\frac{dx^\nu}{d\tau}\right), \tag{16}$$

where is the charge of a mass, and is a electromagnetic field tenor.

With the Einstein's field equation of gravitation given by

$$R_{\mu\nu} = -\frac{8\pi G}{c^4}\left(T_{\mu\nu} - \frac{1}{2}g_{\mu\nu}T\right). \tag{17}$$

From which, Harris obtain equations for the case when the particle is slowly moving compared with the light speed and the gravitational field is sufficiently weak that nonlinear terms in Einstein's field equations can be neglected.

$$\nabla \cdot g = -4\pi G\rho\,, \tag{18}$$

$$\nabla \cdot K = 0\,, \tag{19}$$

$$\nabla \times g = 0 \;\left(\nabla \times g = -\frac{\partial K}{\partial t}\right), \tag{20}$$

$$\nabla \times K = -\left(\frac{4\pi}{c}\right)\cdot(4G\rho v) + \frac{4}{c}\frac{\partial g}{\partial t}, \tag{21}$$

$$F = e\left(E + \frac{v \times B}{c}\right) + m\left(g + \frac{v \times K}{c}\right) + \frac{mv}{2c}\frac{\partial \varphi}{\partial t}, \tag{22}$$

where is a gravitational constant, is a gravitational field, is a co-gravitational field or gravitomagnetic field, is a mass density and is a scalar potential of electromagnetic field. This is similar to the magnetic field in the frame of the electromagnetic theory.

Analogous to Harris's equations, following equations were also given by Heaviside and Jefimenko [22], those are given by

$$\nabla \cdot g = -4\pi G\rho \,, \tag{23}$$

$$\nabla \cdot K = 0 \,, \tag{24}$$

$$\nabla \times g = -\frac{\partial K}{\partial t} \,, \tag{25}$$

$$\nabla \times K = -\frac{4\pi G}{c^2} J + \frac{1}{c^2}\frac{\partial g}{\partial t} \,, \tag{26}$$

$$F = m(v \times K) \,. \tag{27}$$

By using the analogy with the electromagnetism, Forward has shown in his paper that it is possible to built a machine to create a gravitational field by a system of accelerated masses [23,24].

Time varying magnetic field creates a dipole field, the value of the electric field at the center of the torus is given by

$$E = -\dot{B} = \frac{d}{dt}\frac{\mu NIr^2}{4\pi R^2} \,, \tag{28}$$

where is the radius of the torus, is the radius of one of the loops of wire wound around it, and is the total number of turns.

If we replace the wires with pipes carrying a massive liquid, then the known analogy between the electromagnetic and gravitational field can be applied. Then the equivalent gravitational quantities can be given by

$$g = -\dot{K} = -\frac{d}{dt}\left(\frac{\eta NTr^2}{4\pi R^2}\right), \tag{29}$$

where is the gravitational field generated by the total accelerated mass as shown in Figure 4.

In this figure, it is shown the system whose flow is similar to the current flow in a wire-wound ring or torus.

The mass flow through a pipe wound around a torus, which causes a co-gravitational field in a torus. If mass flow is accelerated, the co-gravitational field increases with time and the dipole gravitational field is created. If the pipe is filled with massive liquid and the liquid is moved back and forth in the pipe rapidly enough, then an alternating push-pull gravity field will be generated at the center of the ring. If the machine was big enough and the liquid was dense enough and moving fast enough, we have a gravity catapult that could launch and retrieve spaceships by its gravity repulsion and attraction.

This machine is in the form of a ring of ultradense matter and the flying body is expelled out the other side of the hole with a greatly increased velocity.

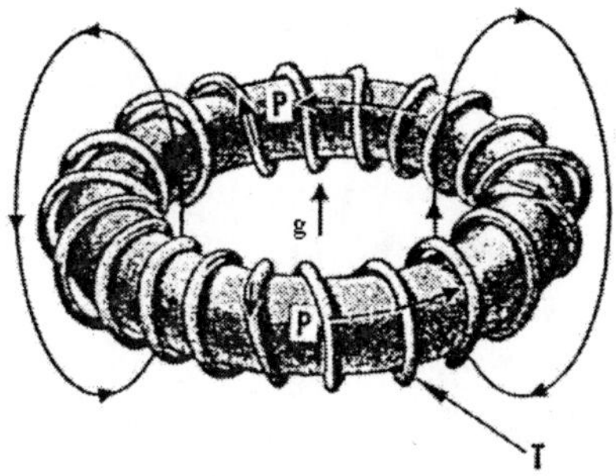

Figure 4. Generator of a Dipole gravitational field.

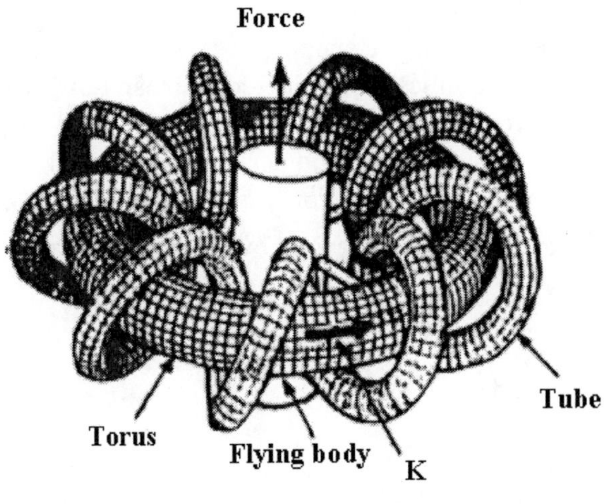

Figure 5. Gravitational machine according to Forward's idea.

Supposing that the gravitational permeability has the value of $\mu = 3.73\times10^{-26}$ m/kg , we have (a : amount of acceleration) at the center of the torus as shown in Figure 4, for the case when the density of a dwarf star through pipes wide as a football field wound around it with kilometer dimensions according to Forward, which is far from present capabilities.

By utilizing this gravitational machine, constant upward gravitational field can be generated which can be used as a gravity catapult that pushes a body as shown in Figure 5.

5.2. Gravitational Machine by Utilizing the Electromagneto-Toroidal Structure

As early as in 1951, P.A.M. Dirac published two papers where he pointed out that we should take into account quantum fluctuations in the flow of the aether [25,26]. Inspired by the Dirac ideas, K.P. Sinha, C. Sivaram and E. C. G. Sudarshan published in 1975 a series of papers that suggested a new model for the aether according to which it is a superfluid state of fermion and anti-fermion pairs, describable by a macroscopic wave function [27-29]. In their papers, they decided to treat the superfluid as the relativistic matter - by putting it into the stress-energy tensor of the Einstein field equations. This allows us to make an important step – to describe the relativistic gravity as one of the small fluctuations of the superfluid vacuum as well. Based on their ideas, we can consider the possibility of space propulsion by utilizing the cogravitational field from the standpoint of hydrodynamics, such a term implies various fluid dynamics which includes linear flow, separated flow as well as a creation of vortices.

At first, we consider the gravitational theory proposed by Heaviside and Jefimenko given by Eqs.(23)-(27). For a quasi-static field, Eq.(26) becomes $rotK \approx -(4\pi G/c^2)J_g$. By replacing by the electromagnetic momentum generated by Poynting's vector S as shown in Fig.5, we have

$$\int_C K \cdot ds \approx \frac{4\pi G}{c^2} \iint_S J_e \cdot dS \,. \tag{30}$$

From the equation $E = V/(r \cdot \log(b/a))$, where is the applied voltage, and are the inner and outer radii of the cylinder shown in Fig.6, we can obtain [30]

$$K \approx \frac{4\pi\varepsilon_0 G}{c^2} V \cdot B, \tag{31}$$

if we suppose there is no cogravitational field outside the tube.

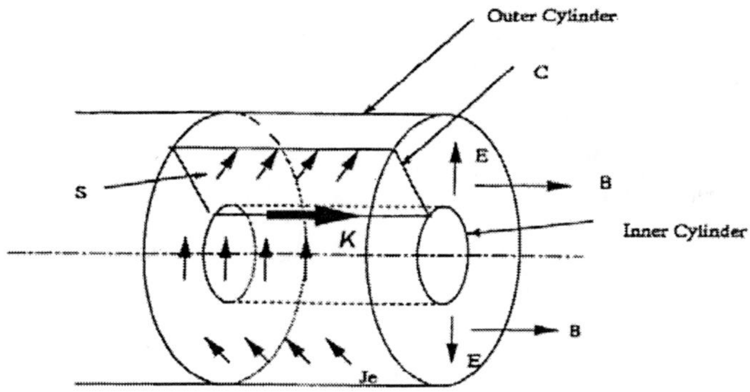

Figure 6. Generation of K by the electromagnetic fields inside the tube structure.

We consider the torus-shaped structure composed of the co-axial condenser and coils rolling around the outer surface of the torus shown as follows:

As the cogravitational field is similar to the vorticity of the fluid dynamics generated inside the torus, shown in Figure 7, this torus-shaped bundle of vortex lines causes the forward motion of the ring, and we may think that the forward thrust for the torus-shaped structure may be generated similar to the fluid dynamics.

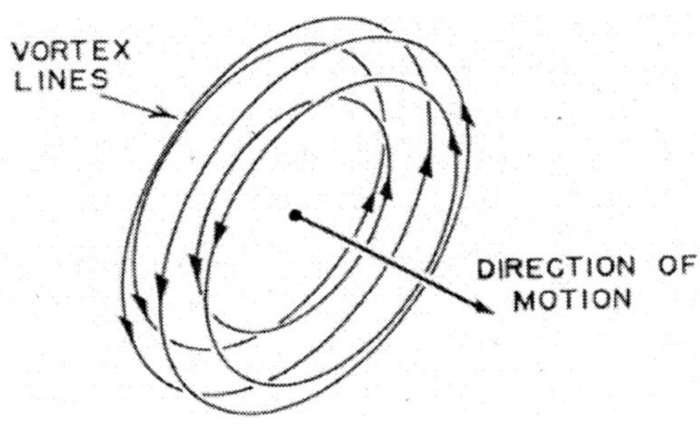

Figure 7. Moving vortex ring in a fluid.

As the electrical toroidal devise is a kind of fluid accelerator, as shown [31]

$$f \cong \frac{\varepsilon_r \mu_r}{c^2} \frac{\partial}{\partial t}[E \times H], \tag{32}$$

we have

$$f \cong \varepsilon_r \varepsilon_0 \frac{\partial}{\partial t}[E \times B], \tag{33}$$

which is shown in Figure 8.

By corresponding electromagnetic and gravitational symbols and a constant given by $E \rightarrow g$, and $\varepsilon_0 \rightarrow -1/4\pi G$ [22], we have

$$f \approx -\frac{1}{4\pi G} \frac{\partial}{\partial t}[g \times K]. \tag{34}$$

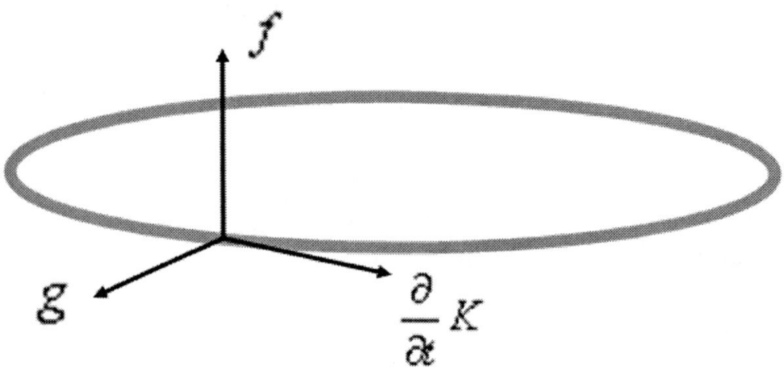

Figure 8. Generated force analogous to the fluid dynamics.

When we let (r ; radius of the toroidal structure, ω: radial frequency of the spinning structure), we have

$$|f| \approx \frac{\varepsilon_0}{c^2} r\omega^2 \left(V \frac{\partial}{\partial t} B + B \frac{\partial}{\partial t} V \right), \tag{35}$$

If $B = const$, we have

$$|f| \approx 9.8 \times 10^{-29} \frac{v^2}{r} B \frac{\partial}{\partial t} V , \tag{36}$$

where is a speed of the rotating toroidal structure.

From the above framework there is the possibility to generate a new gravitational field by applying a impulsive high potential voltage to the toroidal device, as shown in Figure 9.

If we can attain that the velocity of the rotating toroidal structure by one percent of the light speed and we let $r = 5m$, B=100 Tesla, $V = 1$ Giga volt and the duration of time for applying impulsive high voltage electric field to be 1.0 picosecond, we generate a force with magnitude $f = 1.8 \times 10^7$ N, which is sufficiently large to propel a space craft. Hence, it is considered reasonable to envisage the construction of an antigravity machine working as follows:

It is composed of the co-axial condenser which surface is winded by the superconductor coil that generated a gravitational vortex around the cylindrical body. Instead of the Forward's gravity machine which utilizes the circulation of a superfluid through a spiral tubing array, this system uses only high intensity electromagnetic field generated by the co-axial condenser and coils to create a gravitational vortex.

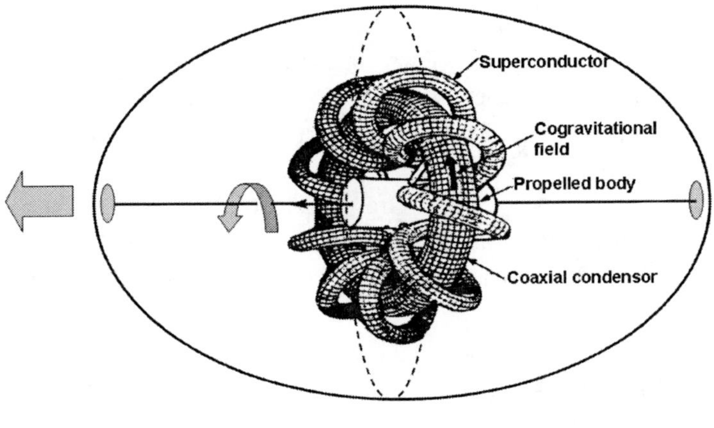

Figure 9. Conceptual design of the propulsion system.

Murad pointed out in his papers that Jefimenko's cogravitational field is the elusive spin or torsion field identified in Russian scientific literature [32, 33]. This idea was first introduced by the French mathematician R. Cartan in

1913, then by Albert Einstein. Within the framework of Cartan-Einstein theory, the existence of these fields has been permitted. Scientists today are recognizing that "spinning fields" really do exist. Just as electromagnetic fields are caused by a charge and gravitational fields are caused by weight, torsion fields are created by any rotating object. Murad and his co-workers reported the rotating devise based upon the Searl and the Russian Godin and Roschin devices [34]. During their test series, they observed weight reduction of 7.3 % at a steady state and a transient lose of 12% of the total weight. They claimed that this device could be applied for developing an advanced propulsion system.

CONCLUSION

In this chapter, some examples of space propulsion methods were advanced, which are in realms that stretch the boundaries of relativity and quantum theories. We may find a way to the stars in the future, deepening our understanding from unusual approaches based on conventional physics, and proposing new directions for space sciences.

REFERENCES

[1] Podkletnov E. Nieminen, R, A possible gravitational force shielding by bulk $YBa_2Cu_3O_{7-x}$ superconductor, *Physica C*, 203; 1992, 441-444.

[2] Wilson J., NASA's Antigravity Machine, *Popular Mechanics*, Dec. 1977, 44-45.

[3] Torr D.G., Li N., Graviotoelectric-Electric Coupling via Superconductivity, *Foundations of Physics*, Vol.6, No.4; 1993, 371-383.

[4] Li N., Torr D.G., Gravitational effects on the magnetic attenuation of superconductors, *Physical Review B*, Vol.46, No.9 ; 1992, 5489-5495.

[5] Li N., Noever D., Robertson T., Koczor R., Brantley W., Static test for a gravitational force coupled to type II YBCO superconductor, *Physica C*, 281; 1997, 260-267.

[6] Modanese, G., Updating the theoretical analysis of the weak gravitational shielding experiments, *Proceedings of the 1997 IAF Congress*, IAA-97, 1997.

[7] Modanese, G., Theoretical analysis of a reported weak-gravitational-shielding effect, *Europhysics Letters*, 35(6); 1996, 413-418.

[8] Modanese, G., Role of a "local" cosmological constant in Euclidian quantum gravity, *Physical Review D*, 54(8); 1996, 5002-5009.

[9] Allen, J.E., Quest for novel force: a possible revolution in aerospace, *Progress in Aerospace Sciences*, 39; 2003, 1-60.

[10] Woods, R.C., Review of claims of interaction between gravitation and high-temperature superconductors, *AIP Conference Proceedings*, 669(1); 2004, 1085-1092.

[11] Podkletnov, E., Modanese, G., Investigation of high voltage discharges in low pressure gases through large ceramic superconducting electrodes, *Journal of Low Temperature Physics*, 132(3-4); 2003, 239-259.

[12] Graham G.M., Lahoz D.G., Obsrvation of static electromagnetic angular momentum in vacuo, *Nature,* Vol.285, No.15; 1980, 154-155.

[13] Lahoz D.G., Graham G.M., Measurement of forces related to electromagnetic momentum in media at low frequencies, *Can. J. Phys.,* Vol.57; 1979, 667-676.

[14] Lahoz D.G., Graham G.M., Observation of Electromagnetic Angular Momentum within Magnetite, *Physical review Letters*, Vol.42, No.17; 1979, 1137-1140.

[15] Corum J.F., Dering, J.P, Pesavento P., Donne A., EM Stress-Tensor Space Drive, *Proceedings of Space Technology and Applications International Forum-1999,* American Institute of Physics,1999, 1027-1032.

[16] Robertson G.A., Murad P.A., Davis E, *New Frontiers in Space Propulsion Sciences, Energy Conversion and Management*, Vol.49; 2008, 436-452.

[17] Slepian J., Electromagnetic Space-Ship, *Electrical Engineering*, Vol.68; 1949, 145-149.

[18] Droscher W., Hauser J., Guidline for a Space propulsion devise based on Heim's Quantum Theory, AIAA 2004-3700, AIAA/ASME/SAE/ASE, Joint Propulsion Conference & Exibit., Frorida, July, 2004, 28-56.

[19] Droscher W., Hauser J., Heim Quantum Theory for Space Propulsion Physics, Space Technology and Applications International From-STAIF, 2005:1430-1440., also available from: www.hpcc.space/ publications/ documents/heim_staif2005-letter.pdf.

[20] Harris E.G., Analogy between general relativity and electromagnetism for slowly moving particle in weak gravitational fields, *American Journal of Physics*, 59 (5); 1991, 421-425.

[21] Braginsky, V.B., Caves, C.M., Thorne, K.S., Laboratory experiments to test relativistic gravity, *Physical Review D,* Vol.15, No.8; 1977, 2047-2068.

[22] Jefimenko O.D., Causality, *Electromagnetic Induction and Gravitation, Electric Scientific Company*, Star City, USA, 1992.

[23] Forward R.L., Antigravity, *Proceedings of the IRE*; 1961, 1442.

[24] Forward R.L., Guidelines to Antigravity, *American Journal of Physics,* 37; 1963, 166-170.

[25] Dirac. P.A.M., Is there an Aether?, *Nature,* vol..168; 1951, .906-907.

[26] Dirac. P.A.M., Is there an Aether?, *Nature*, vol.169; 1952, 702.

[27] Sinha, K.P., Sivaram, C., Sudarshan, E.C.G., Aether as a superfluid state of particle-antiparticle pairs, *Found. Phys.*, Vol.6, No.1; 1976, .65-70.

[28] Sinha, K.P., Sivaram, C., Sudarshan, E.C.G., The superfluid vacuum state, time-varying cosmological constant and nonsingular cosmological models, *Found. Phys.*, Vol.6,No.6; 1976, 717-726.

[29] Sinha, K.P., Sudarshan, E.C.G., The superfluid as a source of all interactions, *Found. Phys.*, vol.8; 1978, .823-831.

[30] Musha, T., Possibility of Antigravity Propulsion by Gravitational Vortex, *Proceedings of the 36th Conference on Aerospace Propulsion, JSASS*; 1996, .321-326.

[31] Pinheiro, M.J., Electromagnetotoroid Structure in Propulsion and Astrophysics, *DFIST*, 2009, 1-7.

[32] Murad, P.A., Its all gravity, Space Technology and Applications International Forum (STAIF-2003). AIP conference proceedings 654, NY, *American Institute of Physics,* 2003, .932-939.

[33] Murad, P.A., A proposed model for capturing gravitational spin anomalies, Space Technology and Application International Forum (STAIF-2004), *AIP conference proceedings 699,* NY, American Institute of Physics; 2004, 1098-1105.

[34] Murad, P.A., Boardman, M.J., Brandenburg, J.E. and McCabe, J., The morning star energy box- part redux, *Journal of Space Exploration*; 2013, 78-86.

Chapter 5

GRAVITY GENERATED BY ELECTROMAGNETOROID STRUCTURES

Mario J. Pinheiro[*]

Department of Physics, Instituto Superior Tecnico
Av. Rovisco Pais, Lisboa, Portugal

ABSTRACT

We introduce the concept of an electromagnetotoroid in astrophysics, and explore its role in the context of astrophysics and modes of animal locomotion. Based on the framework of fluidic electrodynamics, and the analogies provided by this theoretical model, we show that the Abraham's force term is analogous to the Magnus force, and thus represents the formation of electromagnetic vortex structures in the local spacetime structure. We then investigate the possible use of more exotic arrangement of electric currents, such as the apparently ubiquitous toroidal field structure, to provide force to spaceship propulsion. Topological massive theories, used in the context of 3 dimensions, may possibly provide a new framework to study the interaction between matter and fields.

[*] E-mail address: mpinheiro@ist.utl.pt.

INTRODUCTION

The attempt to unify apparently distinct realms of physical phenomena has been preceded by Wilhelm Wien, Max Abraham, Joseph John Thomson, among others, which proposed different kinds of models of the electron and its mechanical properties based on the set of Maxwell's equations, or even proposing appropriated modifications of the orthodox theory of the electromagnetic field [1, 2]. The same basic laws rule over from the microscopic to macroscopic realm, from the laboratory to the interstellar and intergalactic plasmas, and from this vision emerges the plasma universe model, initially sustained by Alfvén and Fälthammar [3] (and discussed in Ch. VI).

Kronenberg *et al.* [3] suggested that the magnetic field lines of galaxies extend a few million light years into the intergalactic medium. Although the mechanism it is not still fully understood, black hole accretion disk energy could be converted into magnetic fields through an efficient dynamo (within charged black holes), a kind of cosmic electric motor. Occasionally, accretion disk systems eject huge amounts of gas (see Figure 1).

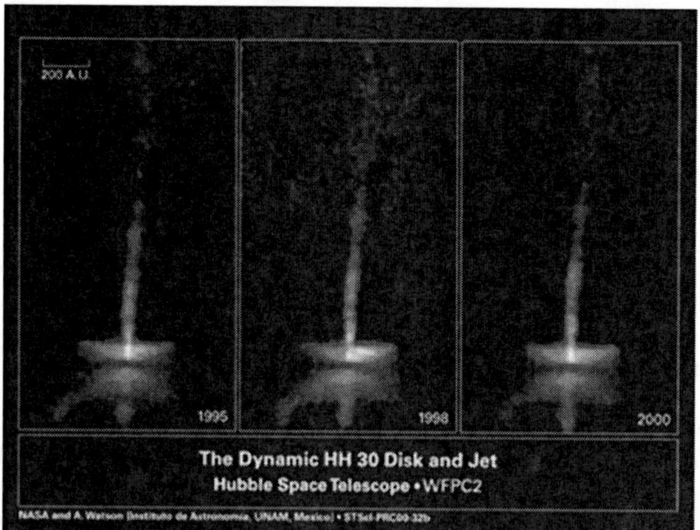

Figure 1. Hubble Space Telescope image of the accretion disk of a forming star, a protostellar object named HH-30, about 450 light years from us in the constellation Taurus. The forming star is occulted in the densest parts of the disk. Courtesy Chris Burrows (ESA).

For example, certain Active Galactic Nucleus expel jets of plasma into space, a phenomena first observed by Alan Marscher's team from the object *BL Lacertae*. The plasma jet from this system spirals outward from a flattened disk of spinning gas surrounding a supermassive black hole, and extends 950 million light years beyond the host galaxy [4, 5].

The bipolar jets appearing in newly formed stars ("pre-T-Tauri") are known as Herbig-Haro (HH) objects (Figure 1). The physics behind the production, acceleration and collimation of relativistic jets from accreting black holes have progressed along two different views: i) jets are the innermost and most energetic region that takes energy from the disk (Lovelace [6]); ii) the jet takes its power from the free energy available due to the spin effect transmitted by the in-fall to the black hole (Blandford & Znajek [7]). According to MHD simulations, magneto-centrifugal ejection may be the driving mechanism [8]. Optical observations [9] indicate that jets are produced in regions 5.5 AU in diameter, while attaining distances of 800 AU from the source. This geometry is a typical Mach angle for free lateral expansion of a supersonic jet [5].

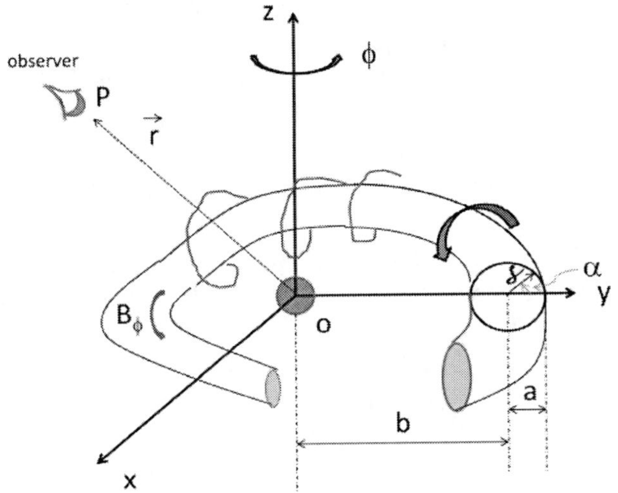

Figure 2. The torus-like geometry that is usually considered when dealing with the infall ionized gas forming the accretion disk of a protostellar structure. Current with density K wrapped around torus.

The accretion disk can be envisaged as a torus-like beam with the current wrapped up around (in helical form), generating the toroidal magnetic field,

B_ϕ. The torus lies in the (x,y) plane, and has minor radius r and major radius R_0, such that $B_\phi = \mu_0 I/2\pi\rho'$, and $\rho' = b + a\cos\alpha$ (see Figure 2).

Nature evolves in ways that respond to its needs, finding systematically new solutions, according to the Optimal Design Principle. Jack Steele, an American medical doctor working for the US Air Force in 1960, at a meeting at Wright-Patterson Air force Base in Dayton, Ohio, coined the word *Bionics*, for the evolving field of study related to copying, imitating, and learning from biology. Later on, Otto H. Schmitt coined the new term *Biomimetics* in 1969, to denote a new field of research implicated in the merging of science and engineering with the purpose to imitate life. These ideas suggest to us to use the working mechanisms and physical structures observed in living beings and the cosmos, to help us with our technologies.

For example, an apparently similar physical structure to the accretion disks can be found in atmospheric phenomena, on two broad classes of lightning-like flashes: sprites [10] and elves [11]. These short-lived, luminous structures are associated with the convective cells of large thunderstorms, and their temporal evolution occurs on time scales from a fraction of a millisecond to a few hundreds of millisecond, while elves form at about 90 km of altitude and sprites at about 70 km. Brief flashes of light in the stratosphere above thunderstorms were first predicted by C. T. R. Wilson [12]. The smallest sprites, named C sprites, are probably single vertical columns, but they can gather together with downward-branching tendrils, or exhibit upward branching toward the ionosphere. Watanabe [13] has presented optical data supporting the conclusion that “column-sprites” are always preceded by elves. They result from the global circuit consisting in a current flow from the top of a thunderstorm to the upper atmosphere, and as well, downward currents to the Earth.

Figure 3 shows the global electric circuit, that takes into account not only magnetic fields, but also currents and electric fields that can be used to understand the electromagnetic conditions in the near Earth plasma [14].

Figure 4 shows one of the first images ever recorded of a blue jet blasting upward from the top of a thunderstorm with a speed in the range 20 to 100 km per sec. Elves are red in color, and appear as a rapidly toroidal disk formed in the sequence of an Electromagnetic Pulse (EMP) from a CG lightning discharge that results from the current change di/dt, with amplitude decreasing inversely with distance according to $1/r$ dependency, as in the far-field of an antenna.

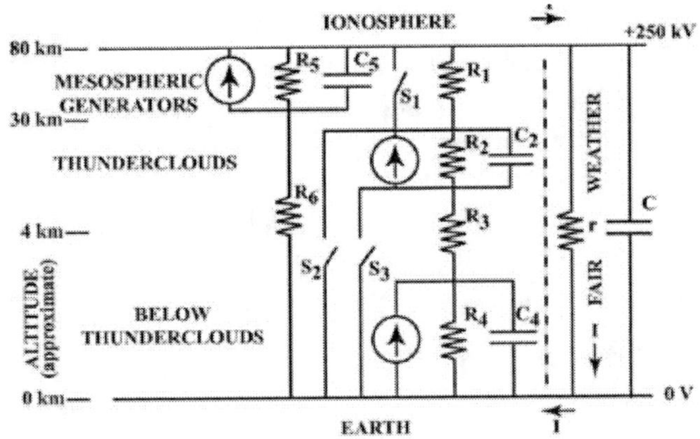

Figure 3. The global electric circuit is a phenomenological model that attempts to explain some relevant properties of sprites, elves and intense lightning discharges. At the left side is the altitude, and at right side the respective fall of potential. The equivalent resistance r represents the fair weather region, and the total current flowing is about 1 kA. As in a RC-circuit, the time constant is $\tau = rC$, on the order of a few minutes. When an upward current occurs, let us say, when a cloud-to-ground discharge takes place, the S_2 switch for that storm closes for about 1 ms, and powers the above circuit, that may trigger a sprite when the switch S_1 closes for a few ms. Reproduced with permission from Ref. [14].

Figure 4. This image was one the first recorded aboard an airplane with a black-and-white television camera, captured over eastern Arkansas at 03:03 UT on UT date July 1. It is a large blue jet, shooting upwards from a thunderstorm's top with a speed in the range of 20 to 100 km per sec, reaching an upper altitude of about 40 km. Image credit: Geophysical Institute, University of Alaska Fairbanks.

These spherically expanding EMP fields are not strong enough to cause breakdown of the atmosphere at low altitudes, and that's why they occur in the region at the bottom of the (planar) lower ionosphere. Also, the 100 μs duration of elves is of the same order of the duration of the high current regime of the lightning strokes [14]. The structure points to a toroidal disk (intersection of the expanding sphere with a plane) through which the sprites emerge [13] (although elves may occur alone [15]). The red spectral observations of sprites are due to the molecular nitrogen electronic transitions $N_2\left(B^3\Pi_g \rightarrow A^3\Sigma_u^+\right)$ with a quenching altitude at about 53 km [15], while the primary blue N_2 emissions identified in the observed spectra is the $N_2(2PG)$ neutral emission from the transition $N_2\left(C^3\Pi_u \rightarrow B^3\Pi_g\right)$.

In biology, several creatures use a propulsion mechanism that relies on the production of vortices by wings, paddles and fins [16]. Fish swim by flapping their tail and other fins, creating vortices in the water that carry away momentum. Squid and salps move by ejecting fluid intermittently, producing vortex rings, with the shape that gives the maximum thrust for a given energy input [17]. Quite interestingly, all these creatures shape their path through the water while the vortex produced at each stroke go behind them, like a motion sustained by traveling through a channel of vortices, as illustrated in Figure 5.

But this new understanding of the problem results from detailed observations that have shown the incredible ability of fishes to reverse direction without slowing down and describing a trajectory with a turning radius only 10 to 30 percent of the length of their bodies, largely overcoming the actual maneuverability of ships, that need to reduce their speed by more than 50 percent, and turn with a radius at least 10 times larger than their typical length [18].

This issue is related to the so-called Gray's paradox, a finding from the British zoologist James Gray in 1936 that after calculating the power needed to propel a dolphin, he concluded that the dolphin was unable by a factor of seven to attain such speeds as 18 ms^{-1}. It should be present another flow mechanism around the blunt body of a dolphin that reduced its drag by a factor of seven, a mechanism besides the only action of the tail that seems so disappointingly inefficient [18].

The careful observation how Nature works, led the non-conformist Viktor Schauberger to state: "The most natural is always the most technically perfect and the only system economically viable long-term" [21].

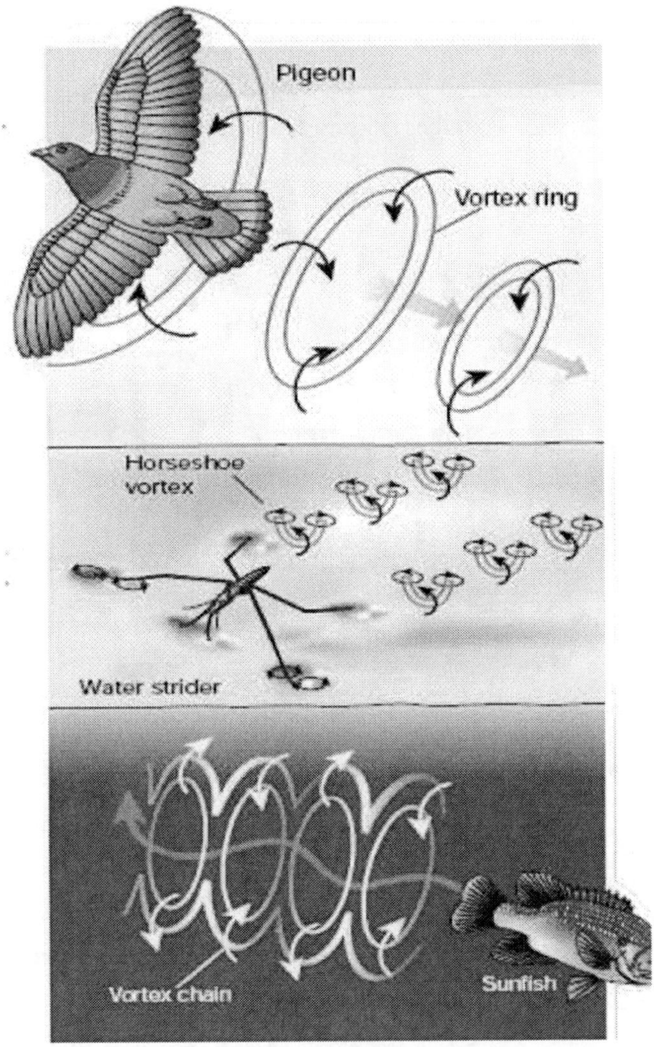

Figure 5. Locomotion in fluids and vortical structure generated at each stroke by bird, water stride and fish. Reprinted with permission from Michael Dickinson, Nature **424** 621-622. Copyright (2003) by the "Nature" magazine (MacMillan Publishers).

Schauberger criticized the tendency that Humanity has to «relate everything to itself (anthropocentrism)». He observed carefully how trout is able to hold its position stationary, where the water is coldest and hence densest, and how its gills could act like vanes able to direct the water flow toward the tail in kind of a vortical counter-current, in this way cancelling the water pressure on the trout's snout. Viktor's vision of science and the natural world led to the outstanding advance in the Nazi aircrafts (Figure 6(b), deepening the modes of swimming like the ray fish (Figure 6(a)), practically no new idea stroke the mind of scientists and engineers, as it is clear with the design of the B2 stealth bomber (Figure 6(a)).

Figure 6. (a) B2-Stealth bomber. Image credit: Northrop Grumman; (b) Horten H IX V-3 (Ho 229) at Silver Hill, Maryland. Image courtesy: Ref.[19]; (c) Bat ray swimming. Image credit: Ref. [20].

The Electromagnetotoroid Structure

In listing the various processes that we have learned about propulsion processes in the natural world, it seems that a particular common principle relies on the production of vortices by a material structure [22] (e.g., wings, paddles, fins).

Maybe this is related to a simple general law of nature, to which is associated we may call a local "twist" of the fluid, firstly investigated by Viktor Schauberger and Professor Von Karman with his experiments with the flow for some distance behind a plate showing a regular arrangement of vortex lines, the so called "vortex street", that remain behind the plate and advancing at a more slow pace than the plate itself, we now turn our attention to the problem of the electrodynamical acceleration of fluids, by reaction against the physical vacuum [23-25], or plasma medium, or any other kind of fluid [22].

Our purpose in this section is to show that Magnus and Abraham forces are a mathematical representation of the same type of structure in the fabric of space-time: a vortex capable of propelling material bodies.

To make the idea of this underlying principle so much simple, we may start to recall that, in the framework of electrodynamics the ponderomotive force acting on the material of an electromagnetic propelling device is provided by Abraham's force density, $\vec{f}_A$, a term which represents the force communicated to a material structure (e.g., Refs. [20,22-29]). It is also known that, in the framework of hydrodynamics, the three-dimensional Magnus force is given by [28]:

$$\vec{f}_{Mxy} = -\rho\left[\left(\vec{V}_v - \vec{V}_{vo}\right)\times\vec{k}_z\right] \tag{1}$$

where is the velocity of the vortex center of mass, is the stream velocity, ρ is the fluid density, and $\vec{k}_z$ is a vector oriented along the z direction with magnitude equal to the circulation. We may start to remark that the "magnetic current force", produced by the magnetic charges that "flow" when a magnetic field changes, is given by $\vec{f}_m = \varepsilon_0\left[\vec{E}\times\left(\dot{\vec{B}} - \mu_0\dot{\vec{H}}\right)\right]$ (see discussion of the idea in Ref. [27]).

This is the "Abraham term" appearing in the Abraham force density $\vec{f}_A$, which differs from the Minkowski force density by means of the expression (see also Refs. [24,29]):

$$\vec{f}_A = \vec{f}_M + \frac{\partial}{\partial t}\left(\frac{\varepsilon_r\mu_r - 1}{c^2}\left[\vec{E}\times\vec{H}\right]\right) \tag{2}$$

The last term of Eq.2 corresponds to the *vacuum-interactance*, a concept introduced in Ref. [30], meaning a process of pushing material bodies through interaction with the physical vacuum, which is associated with the momentum as follows:

$$\vec{g}^V = \frac{\varepsilon_r\mu_r - 1}{c^2}\left[\vec{E}\times\vec{H}\right] \tag{3}$$

But it is well known that a magnetic dipole at rest in an external (and homogeneous) electric field $_{\vec{E}}$ has momentum given by $\left[\vec{\mathscr{m}} \times \vec{E}\right]/c^2$. When the magnetism of the dipole changes, then the density of force is given by the last term of Eq.2.

The analogue hydrodynamic force analogous to Abraham's force is the Magnus force, given by Eq.1. This similarity between vast areas of the physical theory, undoubtedly constitutes a successful example of the unity of physics, and illuminates the physical reality hidden inside the physical vacuum.

In our paper on "fluidic electrodynamics" [22], we introduced a new approach to the realm of the electromagnetic fields, in terms of the potential functions ($\vec{A}, \phi$) and their material derivative, as they emerge in quantum mechanics as more fundamental quantities than the ($\vec{E}, \vec{B}$) fields, predicting certain quantum interference effects, like the Aharonov-Bohm (AB) effect and the single-leg electron interferometer effect, known as the Josephson effect.

The simplified methodology offered by the "fluidic electrodynamics" formulation is a helpful guide to propulsion and energy engineering, avoiding more complicated procedures. There is fundamental reasons to attribute to the vector potential the property of the velocity of a ``fluid'' embedded in the physical vacuum. To describe the inner nature of a so pervasive and evasive medium, be it a mechanical medium whose deformations correspond to the electromagnetic fields, or a locally preferred state of rest, is beyond our purpose now. Possibly this issue is related to the Graham and Lahoz experimental findings that "something in motion" is not duly took into account in our present theories. To describe the electromagnetic field it is necessary to define the electric field $\vec{E}(\vec{r},t)$, the magnetic field $\vec{B}(\vec{r},t)$; charge density $\rho(\vec{r},t)$; and the charge velocity $\vec{v}(\vec{r},t)$. However, it is more advantageous to associate the set of Maxwell's equations with the electromagnetic potentials $\vec{A}(\vec{r},t)$ and $\phi(\vec{r},t)$ through the relationships:

$$\vec{E}(\vec{r},t) = -\vec{\nabla}\phi - \frac{d\vec{A}}{dt} \tag{4}$$

and $\vec{B} = \left[\vec{\nabla} \times \vec{A}\right]$.

Note that we introduced into Eq.4 the convective derivative $d/dt = \partial/\partial t + \vec{v}\cdot\vec{\nabla}$, instead of the Maxwell-Einstein operator $\partial/\partial t$ (see, Ref. [22] and references therein).

Table 1 gives the correspondence of field variables in electromagnetism and hydrodynamics. Notice, for example, that $\vec{B}$ is the analogue to the angular velocity instead of the vorticity.

Table 1. Correspondence of field variables in electromagnetism and hydrodynamics

Electromagnetism	Hydrodynamics
$\frac{q}{\epsilon_0}$	Kraftquelle - Q
Permeability of the vacuum - μ_0	Mass density - ρ
Electric potential - $\phi(\mathbf{r},t)$	Massic enthalpy - $\pi/\rho(\mathbf{r},t)$
Scalar potential - χ	Potential velocity - Φ
Vector potential - $\mathbf{A}(\mathbf{r},t)$	Velocity (or hydrodynamic momentum) - $\mathbf{u}(\mathbf{r},t)$
Electric field - $\mathbf{E}(\mathbf{r},t)$	Lamb vector - $\mathbf{l}(\mathbf{r},t)$
Magnetic field - $\mathbf{B}(\mathbf{r},t)$	**Angular Velocity - 2ω(r,t)**
$U(\mathbf{r},t)$	$p(\mathbf{r},t) = \rho(\mathbf{r},t)\Phi(\mathbf{r},t) - \frac{\rho(\mathbf{r},t)}{2}u(\mathbf{r},t)^2$
Electric current - I	Circulation - Γ
Electromotive force - $\mathbf{E} = -\nabla\phi - \frac{\partial\mathbf{A}}{\partial t} + [\mathbf{v}\times\mathbf{B}] - \nabla(\mathbf{v}\cdot\mathbf{A})$	Hydromotive force - $\mathbf{E_H} = -\frac{\partial\mathbf{u}}{\partial t} - \nabla(\frac{p}{\rho} + \frac{u^2}{2}) - [\omega\times\mathbf{u}]$

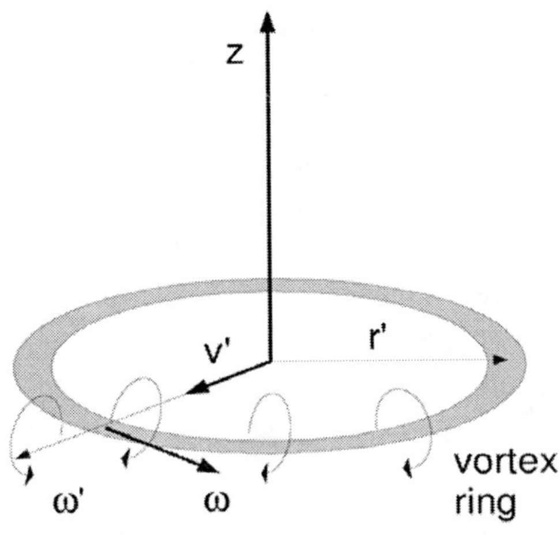

Figure 7. An expanding toroidal disk.

In the next lines, we aim to show how the Abraham's force can propel a plasma jet.

In the natural world, fishes and birds propel themselves through a liquid medium by using their limbs to transfer momentum to the liquid via vortex structures. Their basic mechanism is to form a vortex structure (e.g.: Ref. [21])

analogous to that depicted in Figure 5. The Magnus force is given by (by unit of length), the force vector being perpendicular to both and to the vortex "eye".

We will now look at the mechanism that generates the Magnus force. First of all, we may remark that it is necessary to consume energy in order to progressively enlarge the vortex with a characteristic radial velocity (see Figure 9). The toroidal structure is a vortex ring formed by a closed vortex tube of a given diameter (let us say χ). As is well-known in fluid dynamics, this structure is very stable.

The duty mechanism that provides this radial velocity (inward or outward from the central axis) may have different sources. One example is gas falling into stars, in the case of polar jets; another is the sharp increase of electric current generated by the growing magnetic field of the plasma.

With this kind of mechanism, we can associate a given circulation Γ'(due eventually to an induced field ω'). The falling (or expelled) stream of particles, most probably will acquire a curved trajectory and angular momentum, all effects concurring to the formation of the ring with circulation Γ (and vorticity ω).

At the core of the vortex structure, the resultant force is aligned along the Z axis. Newton's third law predicts a mechanical reaction force $\vec{F}_{mec}$, which can propel a device (or a magnetized fluid). Therefore, we must have (in Figure 7, represents the mechanical force pointing downward).

Now, we intend to show that there is an electrodynamical counterpart - Abraham force - which plays an analogous in the formation of an electromagnetoroid. Let us now explore the concept in more detail. Firstly, replace the ``hydrodynamic magnetization" term in Eq.3 with the *constitutive relationship* $\vec{M} = \eta\vec{\omega}$, where η represents a given property of the medium (a dimensionless constant). The outlined mapping shown in Table 1 gives us the *analogous* hydrodynamic force (by unit of length):

$$d\vec{F}_m^H = -\frac{\rho}{c^2}\frac{\partial}{\partial t}\left[\vec{\omega} \times \vec{l}\right] dv \tag{5}$$

Here, ρ is the mass density and is the differential volume element. We must understand Eq.3 as representing the interaction of magnitudes fed by *different* energy sources: the circulation is associated with motion around the vortex-ring, while the Lamb vector is associated with the increasing vortex radius. The axial vector ω spirals about the azimuthal direction, forming a closed circular loop around the main axis.

It is interesting to note that Eq.4 points to the existence of dual forces: one dependent on the fluid angular acceleration (or time-dependent magnetic force); the other dependent on the Lamb-vector time dependency (or time-dependent electric field).

Let us use the cylindrical geometry, shown in Figure 7, with $\vec{\omega} = \omega_\theta \vec{u}_\theta$ and $\vec{l} = l_r \vec{u}_r$. The total force resulting from this geometry is given by the following expression:

$$\vec{F}_m^H = \frac{\rho}{c^2} \iiint_{S'} \omega_\theta \frac{\partial l_r}{\partial t} \vec{u}_z drdzrd\theta \tag{6}$$

We can arrange terms to obtain

$$F_m^H = \rho \int \omega_\theta \oiint_{S'} \left[\vec{\nabla} \times \vec{\omega}' \right] dzrd\theta dr \tag{7}$$

where we have used the hydrodynamic form of Ampère's equation:

$$\frac{\partial \vec{l}}{\partial t} = -c^2 \left[\vec{\nabla} \times \vec{\omega}' \right] \tag{8}$$

ω' now represents a different (axial) vector (than ω). In fact, it is the vorticity associated with the increasing Lamb vector. The vorticity vector is oriented along the radial axis. C is a characteristic speed of the medium (see Figure 1). Hence:

$$F_m^H = -\rho \int \omega_\theta \oint_\gamma (\vec{\omega} \cdot d\vec{p}) dr \tag{9}$$

However, we also have $\oint_\gamma (\vec{\omega}' \cdot d\vec{p}) = v_r'$. We therefore obtain

$$F_m^H = -\rho \int \omega_\theta v_r' dr = -\rho v_r' \int \omega_\theta dr \tag{10}$$

The last integral of Eq.10 is the circulation (by unit of length). This result can be recast in the following form:

$$\vec{F}_m^H = \rho\left[\vec{v}\times\vec{\Gamma}\right] \tag{11}$$

Eq.11 shows that Abraham's force is the electromagnetic analogue of Magnus's force in hydrodynamics (by unit of length).

Therefore, if the analogy is valid, we conclude that Abraham's force represents a kind of vortex structure formed in the physical vacuum. The associated reaction force can propel a material structure through space. From this general mechanism (see, e.g., [31,32]), we can envisage a mode of spaceship propulsion based on generating electromagnetic vortices [33], along with the development of high-current accelerators and thermonuclear devices.

Construction of Exotic Electromagnetic Arrangements

Is it possible to build certain kind of exotic arrangements of currents or electric charges showing unusual properties with interest for propulsion purposes?

We intend here to show that topological gravitation can give us some clues on this important issue, seeking in particular to classical solutions provided by the Chern-Simons action which leads to simple equations of motion of a system matter plus fields, in three dimensions (2 spatial, one time).

S normally flux of fields occurs in a surface, the use of lower dimensional models are expected to be more easy to handle, and also have important applications in every branch of physics. Just to name a few: in quantum mechanics, constant potential in one-dimensional clarified procedures for energy quantization and tunneling; in statistical mechanics, the two-dimensional Ising model is an exactly solvable model with phase transition; in field theory, spontaneous symmetry breaking and solitons, and Schwinger's two dimensional quantum electrodynamics, gave a boost to stimulate insights into higher dimensional models. For all these reasons, a growing interest in developing theories of gravity in fewer than four spacetime dimensions is growing with the expectation to avoid the mathematical complexities typically found in four dimensions, although some common features remain, such as, nontrivial global and topological effects (see, for ex., Ref. [34]).

TOPOLOGICAL GRAVITATION

The Einstein equation is defined as

$$G_{\mu\nu} = R_{\mu\nu} - \frac{1}{2} R g_{\mu\nu} \tag{12}$$

And the Einstein action is given by

$$S = \frac{1}{2\kappa} \int R \sqrt{-g}\, d^4 x \tag{13}$$

where $g = \det\left(g_{\mu\nu}\right)$ and $\kappa \equiv 8\pi G c^{-4}$, with the usual notation of variables, and $d^4 x = dxdydzdt$. Einstein general theory of gravity (which in its orthodox is massless and at the classical limit is the Poisson's equation, therefore, without taking account transient phenomena, a task performed by D' Alembert equations for potentials) can be modified in order to get a propagating, massive spin 2 degree of freedom.

A possible alteration of this situation can be done through the introduction of the Chern-Simons topological mass term, with the action defined by

$$S_{CS} = \frac{k}{4\pi} \int \delta g^{ij} \varepsilon_j^{kl} \left[\nabla_l R_{ik} - \frac{1}{4} g_{il} \nabla_l R \right], \tag{14}$$

where ε_j^{ij} is the totally antisymmetric tensor density, R_{ij} is the Ricci tensor, R is the curvature scalar, and $k = c^3/4G$. The Cotton tensor (or the three dimensional Weyl tensor) is defined by

$$C^{\mu\nu} = \frac{1}{\sqrt{-g}} \varepsilon^{\mu\alpha\beta} \nabla_\alpha \left(R_\beta^\nu - \frac{1}{4} \delta_\beta^\nu R \right), \tag{15}$$

and is a tensor with symmetric, traceless (metric contraction on any pair of indices yields zero) and identically conserved properties, $n\left(n^2 - 4\right)/3$ terms, and μ is a parameter representing the mass of the linearized topologically massive gravity. The Cotton tensor is the variation of the gravitational Chern-Simons action:

$$\frac{\delta S_{CS}}{\delta g^{\mu\nu}} = \frac{k}{4\pi} C^{\mu\nu} \tag{16}$$

and it is the substitute of Weyl tensor in three dimensional spaces [35], describing non-linear effects of gravitation [36].

The vacuum equations are obtained by maximization of the gravitational and Chern-Simons actions:

$$R_{\mu\nu} - \frac{1}{2} g_{\mu\nu} R + \frac{1}{\mu} C_{\mu\nu} = 0 \tag{17}$$

It was shown [37] that the field equations of topologically massive gravity are given by

$$E_\nu^\mu = \sqrt{-g}\, G_\nu^\mu + \frac{1}{\mu} C_\nu^\mu = -\kappa^2 T_\nu^\mu \tag{18}$$

and

$$C_\nu^\mu = \varepsilon^{\mu\alpha\beta} D_\alpha \left(R_{\beta\nu} - \frac{1}{4} g_{\beta\nu} R \right) \cdot \tag{19}$$

The gravitational and inertial masses differ (see Ref. [38]) by the term:

$$m_g = m_I + \mu S_I \,, \tag{20}$$

with

$$S_I = \kappa^{-2} \int \varepsilon^{ij} x^i E_0^j d^2 x . \tag{21}$$

The Einstein and Cotton tensor together result in a connection between mass and spin [39]. Eq.20 results from the definition of angular momentum

$$J^{\mu\nu} \equiv \int_S \Im^{\mu\nu\alpha} d^2 \Sigma_\alpha \tag{22}$$

where

$$\Im^{\mu\nu\alpha} \equiv \left(x^{\mu} - a^{\mu}\right)T^{\mu\alpha} - \left(x^{\nu} - a^{\nu}\right)T^{\nu\alpha}. \tag{23}$$

If is a surface of constant time t , we have instead

$$J^{\mu\nu} = \int_{S} \Im^{\mu\nu 0} dxdydz \cdot \tag{24}$$

ELECTRIC FORCES LEADING TO A CHANGE IN SYSTEM GRAVITATIONAL MASS

Our purpose here is to outline an approach that may account for gravitational mass changes, and now we want to check out if this model is appropriate to take into account these effects, starting by obtaining the well-known solution by means of our new theoretical approach.

The stress-energy tensor is given by

$$\sigma_{ij} = \varepsilon_0 \left(E_i E_j - \frac{1}{2}\delta_{ij} E^2 \right) + \frac{1}{\mu_0}\left(B_i B_j - \frac{1}{2}\delta_{ij} B^2 \right). \tag{25}$$

We have here $T_{\nu}^{\mu} = g^{\mu\nu} T_{\mu\nu}$, with $g_{\mu\nu} = g^{\mu\nu}$. It can be shown [40] that a point charge located at the origin of a Cartesian coordinate system and supported against a weak static homogeneous gravitational field $\vec{g} = -g\vec{i}$ is given by

$$\vec{E}(x, y, 0, t) = \frac{e}{4\pi\left(x^2 + y^2 + z^2\right)^{3/2}} \left[\left(x - \frac{x^2 g}{c^2} - \frac{y^2 g}{2c^2} \right)\vec{u}_x + \left(y - \frac{xyg}{2c^2} \right)\vec{u}_y \right]. \tag{26}$$

For a system of two point charges q_1 and q_2 located at $\left(x = 0, y = \pm l\right)$, we have

$$\vec{E}_1 = \frac{q_1}{4\pi\varepsilon_0}\left[\vec{u}_z \left(\frac{-l^2\alpha}{2} \right) + \vec{u}_y(-l) \right] \tag{27.a}$$

and

$$\vec{E}_2 = \frac{q_2}{4\pi\varepsilon_0}\left[\vec{u}_z\left(\frac{-l^2\alpha}{2}\right)+\vec{u}_y\left(+l\right)\right]. \tag{27.b}$$

If we look upon the mass term, we would have $m=\int d^3xT^{\alpha 0}\left(\vec{x},t\right)$, with dimensions $MLT^{-2}L$. We must calculate now the term

$$S_I = \frac{1}{\kappa^2}\iint\left(\varepsilon^{xy}xE_y^x+\varepsilon^{yx}yE_x^y\right)dxdy \tag{28}$$

for each mass. It results for each massive particle:

$$S_I^{m_i} = \frac{1}{2}\frac{q_1q_2}{4\pi\varepsilon_0}\frac{g}{c^2};\quad i=1,2 \tag{29}$$

and, therefore, from where it results a changing of the gravitational mass:

$$M_f = m_{f1}+m_{f2}+\frac{q_1q_2}{4\pi\varepsilon_0}\frac{g}{lc^2}. \tag{30}$$

Reminding that the term M_f is in units (Force x Mass), we finally obtain

$$M = m_{01}+m_{02}+\frac{q_1q_2}{4\pi\varepsilon_0}\frac{1}{lc^2}. \tag{31}$$

This is a consequence of the Lorentz transformation, performed on the spatial components of the moment density, $T^{0j}=m^{jk}v^k$, with $m^{jk}=\overline{T}^{00}\delta^{jk}+T^{\overline{jk}}$, and $T^{\overline{jk}}$ denoting the components of the stress-energy tensor in the rest frame of the medium, see e.g. Ref. [42].

This relationship means that the total mass-energy registered by an observer includes the rest mass $T^{\overline{00}}V$ and the work done by forces acting across the volume V. Eq. 34 was obtained after applying the principle of equivalence to obtain the electromagnetic fields of two point charges q_1 and q_2 supported against a weak downward gravitational field [39]. This result shows the intricate relationship between inertial and gravitational mass (or, generally speaking, a "field mass"). The above result may be expected in the frame of the special theory of relativity, representing a mass-renormalization process, as pointed out by Brillouin [44], and showing that the potential energy related to the electrostatic interaction also counts for the total mass of physical bodies.

ELECTROMAGNETOROID ARRANGEMENT LEADING TO A CHANGE IN SYSTEM GRAVITATIONAL MASS

Exotic electromagnetic arrangements of currents and charges can possess unusual properties [45]. When folding a solenoid into a torus (obtaining the structure depicted in Figure 8 an approach based on action shown in Eq. 14 points to the possibility to modify the gravitational mass of the electromagnetotoroid structure.

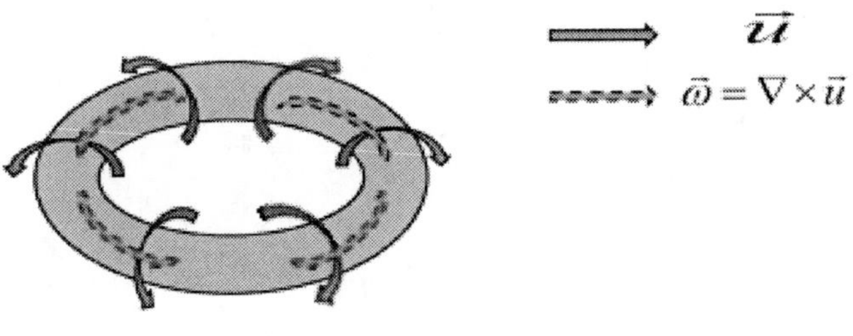

Figure 8. The electromagnetoroid arrangement. The vector $\vec{u}$ represents the fluid velocity and $\vec{\omega}=\vec{\nabla}\times\vec{u}$ its vorticity.

Using simple toroidal coordinates, shown in Figure 2, it can be shown that the current wrapped around the toroidal structure is given by

$$I = K\int_0^r\int_0^{2\pi}(b + a\cos\alpha)drd\phi\,,, \tag{32}$$

and then the density is

$$K = \frac{I}{2\pi r\left(b + \frac{a}{2}\cos\alpha\right)} = \frac{j'}{R_0 + \frac{a}{2}\cos\alpha}, \tag{33}$$

with $j'=I/2\pi a$, and $\vec{J}=K\delta(r-a)\vec{u}_\theta$.

To each turn of the wrapped current is associated an angular momentum, since $\vec{\omega}=\vec{\nabla}\times\vec{u}$. Hence, we write $T_0^i=\varepsilon^{ijk}\partial_j u_k$. Considering that the volumetric current density is given by $\vec{J}=\rho\vec{v}$, we write $u_\theta=K/\rho$. Therefore, we have

$$T_0^i=\frac{1}{\rho}\varepsilon^{ijk}\partial_j\left[K\delta(s-a)\right]. \tag{34}$$

From Eq.18, we obtain

$$E_0^\theta=-\frac{\kappa^2}{\rho}\frac{I}{2\pi a^2}\left[\frac{1+\frac{a}{2}\cos\theta}{b+\frac{a}{2}\cos\theta}\right]. \tag{35}$$

Inserting into Eq.10, we obtain

$$E_0^\theta=-\frac{\kappa^2}{\rho}\frac{I}{2\pi a^2}\int rE_0^\theta ad\theta 2\pi b=-\frac{Ib}{a\rho}\int_0^{2\pi}d\theta\frac{1+\beta\cos\theta}{b+\beta\cos\theta}, \tag{36}$$

which gives approximately the following result

$$S_I\approx-\frac{Ib}{a\rho}2\pi. \tag{37}$$

Thus, the gravitational mass is subject to a change according to the formula:

$$m_g=m_I+\mu S_I\approx m_i-\frac{2\pi Ib}{a^2\rho}. \tag{38}$$

We conclude that the mass variation of the electromagnetotoroid device is directly proportional to the electric current intensity I, to the great radius b, and inversely proportional to the inner radius a and the charge density.

According to the sign of ρ the increment can be positive or negative. We may also remark that Eq.31 is a kind of Gauss's law.

Conclusion

A large number of natural systems seem to be based on the working mechanism of toroidal configurations, or vortices. Therefore, we introduce the concept of an electromagnetotoroid configuration, and explore its role in the context of astrophysics and modes of animal locomotion. The framework of fluidic electrodynamics, and the analogies provided by this theoretical model, suggests that the Abraham's force term is analogous to the Magnus force, and we are therefore tempted to associate this analogy to the formation of electromagnetic vortex objects in the local spacetime structure. We then investigate the possible use of more exotic arrangement of electric currents, such as the apparently ubiquitous toroidal field structure, to give a sketch of possible consequences for spaceship propulsion. Finally, we attempt to apply topologically massive theories, used in the context of 3 dimensions in various fields of physics. This theoretical framework seems useful, since a great number of forces result from the flux of fields through surfaces, and we first check it in the classical problem of variations of mass of charged particles in the presence of the gravitational field, later applying the model to the more exotic arrangement of currents in a toroidal configuration, showing the conditions for changing the gravitational mass. It will require much work in order to gain more understanding, but topologically massive gravity theories may possibly provide a new framework to study the interaction between matter and fields, with utility for advanced spacecraft propulsion systems.

References

[1] Darrigol, *Olivier Electrodynamics from Ampère to Einstein*; Oxford University Press: Oxford, 2002.

[2] Mehra, Jagdish, Einstein, Hilbert, and the Theory of Gravitation-Historical Origins of General Relativity Theory; D. Reidel Publishing Company: Dordrecht-Holland, 1974.

[3] Alfvén, Hannes O. G. *IEEE Trans. Plasma Science* 1999 18 5-10.

[4] Kronenberg, P. P., Dufton Q. W., Li H, Colgate S. A. *Astrophys. J.* 2001, 560, 178-186.
[5] Marscher Alan P. *et al. Nature* 2008 452 966-969.
[6] Lovelace, R. V. E. Nat. 1976, 262, 649.
[7] Blandford, R. D.; Znajek, R. L., MNRAS 1976, 179, 433.
[8] Cabrit, S. *Jets from Young Stars*, Lect. Notes Phys. 2007, 723, 21-53.
[9] Hartigan, P; Edwards, S; Pierson, R., *Ap. J.* 2004, 614, L69.
[10] Sentman, D.D.; Wescott, E.M. *Geophys. Res. Lett.* 1993, 20(24), 2857.
[11] Lyons, W. A. *Geophys. Res. Lett.* 1994 21(10) 875.
[12] Wilson, C. T. R. *Proc. Phys. Soc. London* 1925 37, 32D.
[13] Watanabe, Y. *A Study on Space-time Structures of Sprites based on Photometric Observations,* Master's thesis, Tohoku University, 1999.
[14] Rycroft, Michael J. Introduction to the Physics of Sprites, Elves and Intense Lightning Discharges, Ed. Füllekrug; Mareev, Eugene A.; Rycroft, Michael J. Sprites, elves and Intense Lightning Discharges, Springer: Dordrecht, 2004.
[15] Siskind, David E.; Eckermann, S. D.; Summers, M. E. (Eds.), *Atmospheric Science across the Stratosphere*, Geophys. Monogr. Ser. Vol. 123; AGU, Washington, D. C., 2000.
[16] Dickinson, Michael *Nature* 2003 424 621-622.
[17] Linden, P. F. and Turner, J. S. *2004 Proc. R. Soc. Lond. B* 271, 647–653.
[18] Trianafyllou, Michael S.; Triantafyllou, George S. 1995 *Scientific American* 272 (3), 64.
[19] Dabrowski, H. P. *The Horten Flying Wing in World War II*, Schiffer Military History, West Chester, PA, 1991.
[20] Stone, M., Lynn, *Rays-Marine Life*, Rourke, Florida, 2006.
[21] Viktor Schauberger, *The Water Wizard: The Extraordinary Porperties of Natural Water*, Trans. And Ed. By Callum Coats, Vol.1, Eco-Technology, 1997.
[22] Martins, A. A.; Pinheiro, M. J. 2009 *Phys. Fluids* 21, 097103.
[23] Martins, A. A.; Pinheiro, M. J. 2007 *AIP Conference Proceedings* 880, 1189.
[24] Feigel, A. 2004 *Phys. Rev. Lett.* 92(2) 020404-1.
[25] Martins, A. A.; Pinheiro, M. J. 2008 *Int. J. Theo. Phys.* 47 (10), 2706-2715.
[26] Harpaz, A.; Soker, N. 1998 *Gen. Rel. Grav.* 30 (8), 1217-1227.
[27] Shockley, W.; James, R. P. 1967 *Science,* 156 (3747), 542.
[28] Huggins, Elisha, R. 1970 *Phys. Rev. A* 1 (2) 327-331.
[29] Brito, Hector H. 2004 *Acta Astronautica* 54, 547-558.

[30] Cleveland, Blair, M. 1996 *Electric* Spacecraft 24, 6-16.
[31] Pinheiro, Mario J. 2011 *Physica Scripta* 84 (5), 055004.
[32] Hedenström, Anders 2004 *TRENDS in Ecology and Evolution* 19(5) 217-219.
[33] Robertson, G. A.; Pinheiro, M. J. 2011 *Physics Procedia*, 20, 6-23.
[34] Brown, J. D. Lower Dimensional Gravity (World Scientific, Singapore, 1988).
[35] Deser, S; Jackiw, R.; Templeton, S. 2000 Ann. Phys. 281, 409-449.
[36] Deser, S. *1990 Phys. Rev. Lett.* 64, (6) 611-614
[37] Gregory Hood,C. 1966 *Phys. Rev.* 143 (4), 1006-1011
[38] Deser, S.; Jackiw, R.; Templeton, S. 1982 *Ann. Phys.* (N.Y.) 140, 372
[39] Boyer, T. H. 1979 *Am. J. Phys.* 46(4) 383-385; Boyer, T. H. 1979 *Am. J. Phys.* 47(2) 129.
[40] Choquet-Bruhat, Y. *General Relativity and the Einstein Equations* (Oxford University Press, Oxford, 2009.
[41] Garcia, A. A.; Hehl, F. W.; Heinicke, C.; Macías, A. Preprint at http://arxiv.org/pdf/gr-qc/0309008.pdf
[42] Deser, S.; Tekin, B. 2003 *Class. Quantum Grav.* 20 L259.
[43] Misner, C. W.; Thorne, K. S.; Wheeler, J. A. *Gravitation* (W. H. Freeman, San Francisco, 1973).
[44] Brillouin, L 1965 *Proc. N. A. S.* 53(3), 475-482.
[45] Miller, M. A. 1984 *Sov. Phys. Usp.* 27(1) 69-75.

Chapter 6

CONNECTION BETWEEN THE PLASMA UNIVERSE THEORY AND THE GRAVITO-ELECTROMAGNETIC FORCES

Mario J. Pinheiro[*]
Department of Physics, Instituto Superior Tecnico
Av. Rovisco Pais, Lisboa, Portugal

ABSTRACT

Theoretical plasma physics has been developed along the line of thought provided by the kinetic theory of gases, practically introducing only the features of charged particles. However, a growing number of data provided by astronomical and laboratory observations, lead us to a confrontation that points to a lack of appropriate concepts. The apparent incapability of thermonuclear physics to create a sustainable source of energy is a proof that the actual theoretical frame in plasma physics is incomplete. The newly acquired knowledge, via experiments and theoretical advances, in the areas of magnetospheres and interplanetary plasmas, physics of interstellar clouds and formation of stars, is expected to contribute to a new vision of the universe (the plasma universe) and providing new theoretical tools that will enrich our knowledge and, in particular, might provide new concepts for advanced propulsion.

---

[*] E-mail address: mpinheiro@ist.utl.pt.

INTRODUCTION

If man had no representations, he would not think;
Were he not an imaginative spirit, he would not be a logical spirit.

-Benedetto Croce, in Logic as the Science of the Pure Concept

We live in a plasma dominated universe, and, therefore, the understanding of the processes that occur in this medium is of fundamental importance. It is natural to assume that, in the past, the universe was in a plasma state, which implies that this study also must be considered when dealing with cosmological issues [1]. However, a growing number of experimental data confront scientists with need to revise the field of theoretical plasma physics that was developed along the lines of thought provided by the kinetic theory of gases (e.g., Klimontovich's microscopic formalism and the Bogoliubov-Born-Green-Kirkwood-Yvon hierarchy equations for multiparticle distribution functions), with new concepts that may provide a new framework to understand

The plasma state consists of electrically charged particles that behave collectively under the influence of electromagnetic fields. The charged particles are electrons, positive positive (protons H^+, and N_2^+, O^+, NO^+) and/or negative ions, but they can also be formed by charged grains or dust particles. The density of the plasma is usually low (except inside stars) so that short-range binary collisions are negligible, while long-range electromagnetic forces, predominates. The term *plasma* was given in 1923 by Irving Langmuir, Nobel Prize in chemistry, name taken from the medical term plasma used to designate the medium that fills and shapes to the cell, as he was fascinated by the almost lifelike properties of this medium. The electromagnetic field is well described by the set of Maxwell's equations:

$$\frac{\partial \mathbf{B}}{\partial t} = -\nabla \times \mathbf{E} \tag{1}$$

$$\frac{\partial \mathbf{D}}{\partial t} = \nabla \times \mathbf{H} - \mathbf{J} \tag{2}$$

$$\nabla \cdot \mathbf{D} = \rho \tag{3}$$

$$\nabla \cdot \mathbf{B} = 0 \tag{4}$$

This set of equations are defined when the Lorentz force is given by

$$\frac{d}{dt}(m\mathbf{v}) = q(\mathbf{E} + \mathbf{v}\times\mathbf{B}); \qquad \frac{d\mathbf{r}}{dt} = \mathbf{v} \tag{5}$$

and the constitutive equations are defined by $\mathbf{D} = \varepsilon\mathbf{E}$ and $\mathbf{B} = \mu\mathbf{H}$, relating the electric field vector $\mathbf{E}$ and the displacement vector $\mathbf{D}$, and as well the magnetic flux density vector $\mathbf{B}$ and magnetic field vector $\mathbf{H}$, while ε and μ are the permittivity and permeability of the medium, ρ and $\mathbf{J}$ are the electric charge and current densities, respectively. m and q are the mass and charge of the particle. The word plasma was coined by Irving Langmuir (1881-1957) [2], a term that ensued from his observations of the separation of the plasma into cell-like regions with boundaries formed by charged particles sheathes, whenever regions with different densities, temperatures, or electromagnetic fields inhomogeneity are present.

The plasma state is globally neutral, this meaning that the number of positive charges equals the number of negative charges. If $\overline{n}_e$ represents the average number density of electrons (average number of electrons per unit of volume) and $\overline{n}_z$ is the average number density of ions with ionic positive charge Ze, then the global condition is satisfied:

$$\overline{n}_e = \sum_Z Z\overline{n}_z \,. \tag{6}$$

This is the condition of quasi-neutrality. If, in addition, also negative ions with average density are present, we must modify the above Eq.6, such as:

$$\overline{n}_e + \overline{n}_n = \sum_Z Z\overline{n}_Z \,. \tag{7}$$

The range of values of the plasma density spans from a few electrons per unit of volume up to 10^{20} electrons per cubic centimeter. The temperature range goes from a few milli-electronvolt (eV) registered in neon and fluorescent tubes to thousands eV in stars. The plasma state can be created at the cost of a certain amount of energy, achieved by injecting energy to a gas or vapor in order to ionize the constitutive atoms and molecules. The source can be external, or placed at the heart of the medium through exothermal chemical reactions or nuclear reactions, for instance. After the energy source is disconnected, the natural process is the attraction of positive to negative particles, the plasma recombination process.

The plasma state is characterized by certain properties, such as the Debye length, neutrality, and the so called plasma parameter.

Besides the cellular structures that plasma can form, it is also possible the formation of filamentary structures, since electrons can flow in a background of positive ions, and the resulting current form a ring of magnetic field around it, subject to typical instability, known as pinch effect.

THE PLASMA UNIVERSE THEORY

The idea of the plasma universe model were mostly advanced by Hannes Alfvén (1908-1995) and C. G. Fälthammar [1, 3]. The basic idea sustains that the same physical laws holds from the laboratory to the intergalactic space, where a network of electric currents fill the space, while at the same time transfer momentum and energy over large distance in a kind of cellular structure, see figure 1, which illustrates a kind of neurocosmos framework, also supported by the approximate magnitude of constitutive elements in the two different levels, 100 billion brain cells, which seems close to the 125 billion galaxies estimated by the Hubble telescope of the Deep Field, a number that tends to increase with the astronomical observations with other instruments, such as radio telescopes, infrared cameras, X-ray and γ -ray cameras, opening the spectra well beyond.

Apparently, we have in this fact an argument to support those that believe that the human brain was created at the image of the Cosmo, such as the human being was capable to map the entire universe [4].

Figure 2 shows the spectrum of the cosmic background radiation spectrum. We see a peak at frequency $\nu = 2.79 \times 10^{11}$ Hz, which according to Wien's displacement law $\lambda_m T = 2.898 \times 10^{-3}$ m.K, corresponds to the temperature of the cosmic background radiation. The existence of this peak suggests possible means to extract energy from this universal "sea" of energy [5]. The possibility to extract energy from the "radiant energy", as called by Nikola Tesla, attained nowadays an increasing importance with the global energy crisis, since our civilization is ultimately condemned to collapse.

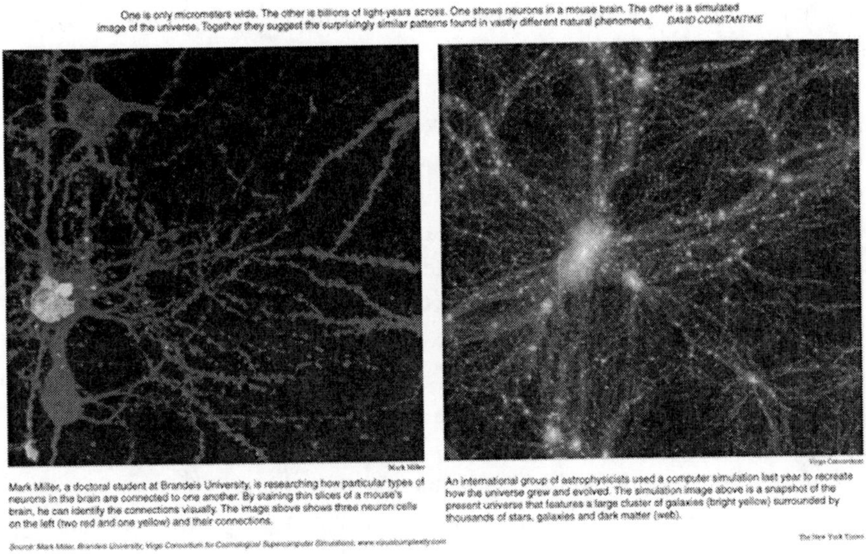

Figure 1. The structure of a brain cell is the same as the entire universe. Courtesy of.Mark Miller, Brandeis University.

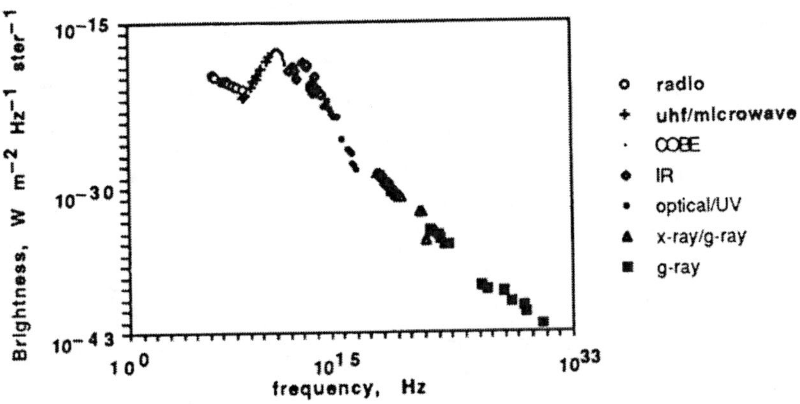

Figure 2. Spectrum of the cosmic background radiation. Notice the peak at about 10^{11} Hz, corresponding to the temperature of the cosmic background radiation.Courtesy of A. Peratt [14].

Michael Faraday (1831) was possibly the first to observe the phenomenon of parametric resonance, as he noticed that surface waves excited by vertical excitation of a fluid-filled column had half the frequency of the frequency of excitation.

The theory of parametric resonance, established in great part by L. I. Mandelstam and N. D. Papaleksi (see, e.g., [6-7]), suggest the possibility to control and amplify fluctuating external driven force acting on a physical system. The resonance phenomena have a particular relevance in the natural world, science and technology, medicine, civil engineering, aeronautic, rocket technics and astronautics, for instance.

Parametric resonance and nonlinear phenomena explain to some extent, the process of self-organization and adaptive grouping in stable formations. The most popular example of a parametric resonance is a child swinging on a swing. The child movements at the extreme points correspond to a periodic change of the physical pendulum length, and with this movement the amplitude increases greatly.

The phenomena of parametric resonance differs from the forced oscillations, because in this last one what counts is the matching of the driving frequency with the oscillator's natural frequency, while the strongest parametric resonance are excited when the modulation's frequency is twice the system's natural frequency ω_0 :

$$\omega = \frac{2\omega_0}{n}, \tag{8}$$

where $n = 1, 2, \ldots$ In a linear system, the amplitude of parametrically excited oscillations can grow indefinitely. However, in real life situations, nonlinearities restrict the growth of the amplitude. In addition, parametric resonance can only take place if the system is already characterized by, at least, some feeble natural oscillations, and with friction, it is necessary that the parameter's amplitude of modulation exceed a certain threshold.

A typical mechanical-electrical system that exhibits parametric resonance is shown below in Figure.

This circuit consists of an LC circuit with a constant inductor L connected in series with a capacitor whose value changes with time due to its electrodes (with charge q) subject to a prescribed motion $d(t)$, and hence $C(t) = \varepsilon S/d(t)$, with the usual notation (and current $i = \dot{q}$). Kirchhoff's laws give immediately

$$\ddot{q}+\frac{d(t)}{\varepsilon SL}q=0 \tag{9}$$

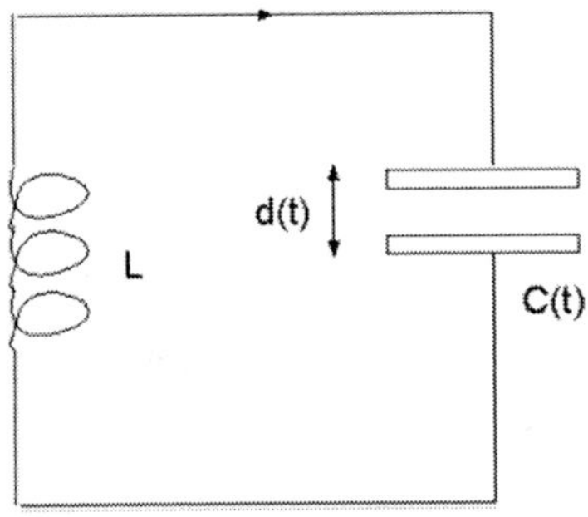

Figure 3. A mechanical-electrical system exhibiting parametric resonance.

The study of parametric resonance in a dynamical system, lead us frequently to the Mathieu's equation, a special case of second order differential equation with a periodic coefficient

$$\frac{d^2u}{dt^2}+(a-2q\cos 2t)u=0 \tag{10}$$

or to the more standard form of the Mathieu-Hill equation

$$\frac{d^2u}{dt^2}+a(t)u=0 \tag{11}$$

with $a(t+T)=a(t)$, T denoting the parametric resonance period of excitation. The general solution of Mathieu's equation, as a consequence of Floquet's theorem, is

$$u(t)=Ae^{\mu t}\phi(t)+Be^{-\mu t}\phi(-t) \tag{12}$$

where A and B are constants of integration, μ is the periodicity exponent and $\phi(t)$ has period π (see, e.g. [8]).

Parametric resonance is at the basis of a new discipline called *Energy Harvesting*, and the usual devices used are micro-electromechanical system

(MEMS) device, made up of a micro-cantilever structure and a transducer. It is enough that an external force applied to the cantilever perpendicularly to the length, instead of transversely, for parametric resonance mechanism to generate more energy from the same amount of vibration, from either the natural vibrations of the environment, or any other source [9].

The more comprehensive theory of resonance in nonlinear oscillating circuits is at the basis of *Synergetics*, or more generally, the theory of self-organizing systems, that aims to explain the formation of typical phenomena, as shown in Figure and as well cosmic formations, as the intergalactic structure illustrated in Figure4.

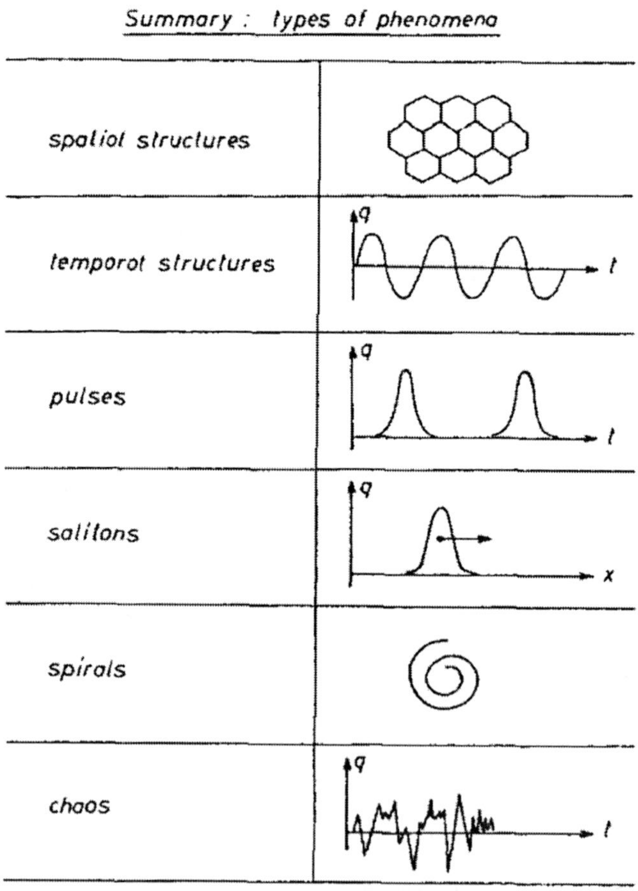

Figure 4. Types of phenomena studied in the frame of synergetics. Courtesy of H. Haken [10].

Which process brings galaxies together from apparently unrelated regions of the Cosmo to form systems with typical structures? Observational cosmology sustains that the so called large structure of our world, i.e., clusters and superclusters, is most probably the outcome of small perturbations of matter and gravitational potential occurred in the primordial gas, in the isotropic Friedmann background [11]. However, this view is not entirely shared by the plasma universe theory.

The cellular structure observed results from the presence of electric and magnetic fields associated with planets, stars, including exotic stellar objects as pulsars and neutron stars. With no doubt, magnetic forces can be locally much higher than gravitational forces, imposing its own dynamical consequences.

On Earth's magnetosphere these fields have been measured directly with instruments carried on spacecraft, but to fully understand how these electromagnetic fields are generated requires to grasp the collective properties of charged particles in the fluid approximation, and as well, the concept of current layers in space. Magnetized planets and celestial bodies exhibit a nearly magnetic dipole (axisymmetric) field, that together with the interplanetary magnetic field (IMF) usually result in phenomena with higher complexity. The IMF has solar origin and it was inferred in sunspots firstly by G. E. Hale in 1908, permeating the space between the Sun and planets. Near the Earth, the IMF has a week value of about 10^{-9} teslas (contrasting with its value at the surface of the Sun, which is about 10^{-4} teslas).

But the inspirational first model of the Sun as the source of the solar corpuscular radiation was given by the Norwegian scientist Kristian Birkeland (1867-1917), when he did his terrella experiment (1896), consisting of a 320 liter vacuum chamber with a hot cathode at the bottom and top surfaces simulating the Sun, and the uniformly magnetized sphere with 24 cm simulating the Earth (the terrella, Latin of "little earth") as the anode (see Figure 5).

An electrostatic potential between the cathode and the anode, creates a stream of electrons (corona discharge) and its impact on a thin region of the polar caps gave a fluorescence that mimic the aurora lights. Birkeland was convinced that Earth and the Sun could be represented as a Crookes tube at a cosmic scale, with the sunspots acting as the cathodes, source of electrons which collide and stream along the geomagnetic field, and giving rise to the aurora [5]. Andres Celsius was the first to register observations of aurora borealis in Sweden, and made a report to the Proceeding of the royal society entitled 'Observations of the Aurora Borealis in Sweden', in 1722. He made observations using a big compass needle, and thus referring for the first time, an electromagnetic phenomena as the source of the aurora.

Plasma fills practically the entire universe (the word meaning unity in diversity), in fact is the dominating state in the universe, with the exception of certain regions, as for example, the Earth's surface and the upper atmosphere, due to the Earth's magnetic field, a protective cocoon that protects our planet, as illustrated in Figure 6.

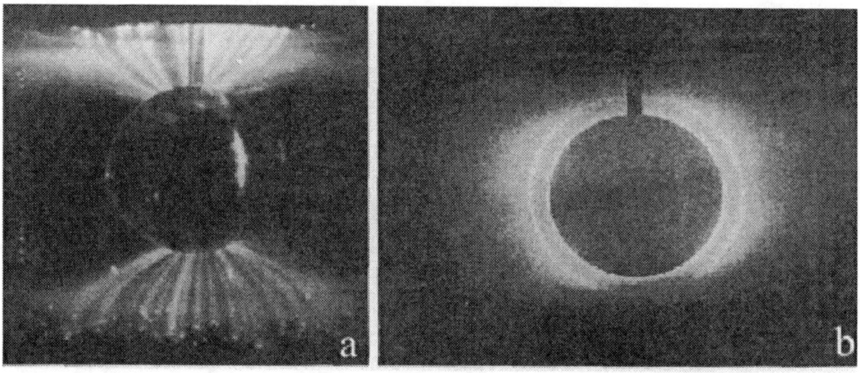

Figure 5. (a) Corona discharges between the sperical terrella (anode) and the surfaces at the top and bottom; (b) Simulation of the Sun's corona with the Birkeland terrella experiment in the presence of a very low magnetic field (Courtesy of J. Phys. IV France).

Before 1950's it was generally assumed that the interplanetary space was pure void, although it was already known that occasionally the sun was the source of a burst of particles when solar flares occurred. In fact, satellite observations have shown that the Sun continuously creates plasma in the interplanetary space.

The source of the solar wind is the lower corona, the charged particles are accelerated radially from the Sun, attaining supersonic speeds at a few solar radii while the plasma becomes more rarefied (collisionless, meaning that the collisional mean-free-path exceeds the typical length for density variations, and resulting in an electric current flow with minor resistance). This situation favor the "frozen in" of the magnetic field lines to the solar wind, which is thus carried away from the Sun into space; this is the interplanetary magnetic field (IMF).

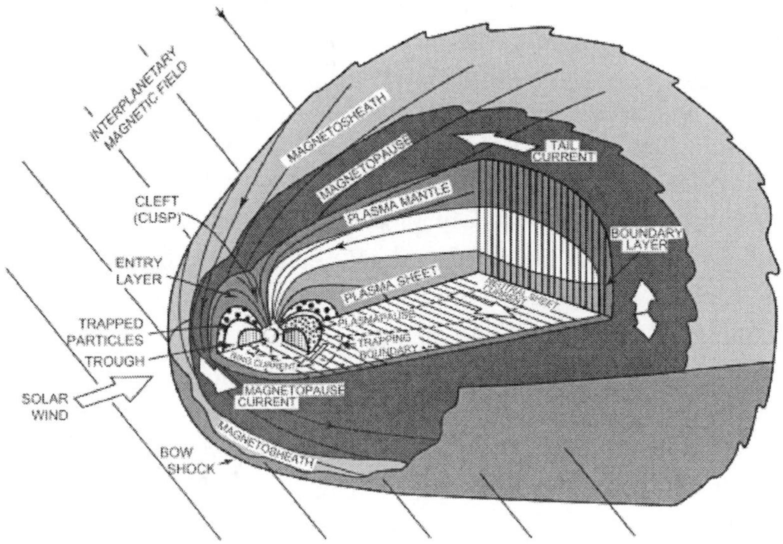

Figure 6. The Earth's magnetoshpere inside the bow shock. Courtesy of Elsevier (taken from the book "Earth's magnetospher formed by the low-latitude boundary layer, by Walter Heikkila).

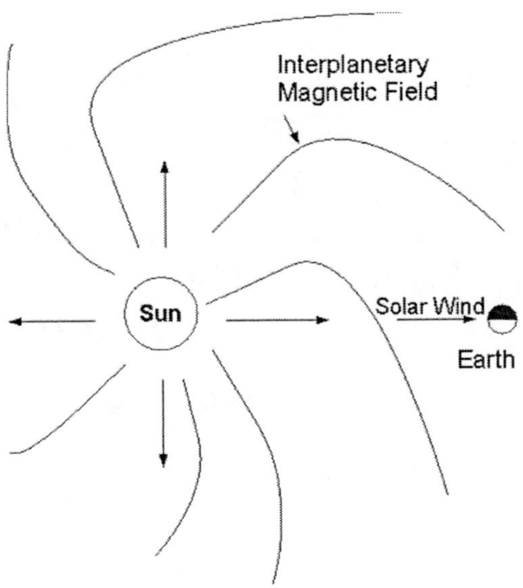

Figure 7. The magnetic field lines are "frozen in" in the solar wind (plasma) that flows radially from the Sun, but due to the slow rotation of the Sun, the magnetic field lines bend into spirals, what is called the interplanetary magnetic field.

The Sun is in a plasma state, is not just a hot gaseous stellar object (expected to be cooling with distance), and the proof is that its temperature increase from its surface in the chromosphere to the corona, where its temperature increases abruptly. Our knowledge of the processes governing the abundance of elements, nuclear reactions, and the formation of the internal magnetic fields are still incomplete. However, the Sun's surface structure is more complete, constituted of a photosphere (a weekly ionized atmosphere $T \approx 0.5$ eV), chromosphere ($T \approx 4$ eV, extending 5,000 km above the transition region to the inner corona), and corona (characterized by plasma temperatures $T \approx 70-263$ eV and the place of explosively unstable magnetic field configurations, solar flares and X-ray emission).

Besides the Sun, in our solar system stands another large plasma structure around Jupiter and Saturn, the Jupiter-Io plasma torus, filled with sulphur ions with density of 3×10^{3} cm^{-3}. A huge weakly ionized hydrogen plasma torus encircles Saturn, extending from an outer diameter 25 times the radius of the planet and inner diameter around 15 Saturn radii (see Figure 8). Connecting Jupiter and Io is a huge Birkeland current carrying about seventy times the total electric power generated by all nations on our planet. No fusion experiment was capable to replicate these configurations.

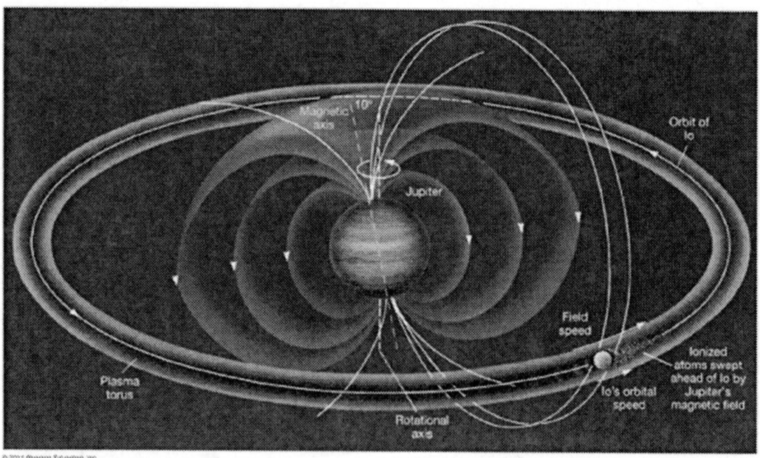

Figure 8. The Jupiter-Io plasma torus and the magaampere Birkeland current flowing between Jupiter and Io. Illustration credit: NASA.

Comets also possess they own "magnetospheres". Figure 10 shows the typical structure that forms due to the interaction of the solar wind with a comet. The nucleus of a comet spans typically along several kilometers across and due to the solar extreme ultraviolet photons, a number of cometary ions (H_2O^+, O^+, OH^+, CH_n^+) are formed through ionization of the neutrals. These ions are created practically at rest with the fast solar wind flow, responding quickly to the electric and magnetic field associated with the solar wind. The momentum transfer between the solar wind and the newly born ions, slow down the solar wind flow that bends the field lines, as shown in Figure 10. If this interaction is strong enough, a bow shock forms around the comet. This is an example of a transition region, a boundary layer that separates regions of plasmas with different densities, temperature, magnetization, and chemical composition [13].

Figure 9. The robotic space probe Pioneer 10 in the main assembly building. It allowed direct observation of the Jupiter's magnetic field. Image credit: Wikipedia.

Charged particles are accelerated in an electromagnetic field according to Eq.5. On Earth, particle accelerators can be operated through the potential gradient between a cathode and anode, like in a cathode ray tube (CRT), pulsed high-voltage, betatrons and resonant cavities and waveguides, among other mechanisms. In 1952, Alfvén and Wernholm suggested the use of an electron beam for an accelerating system [15,16]. The main idea rely on the high fields associated with bunches of particles to drag other particles along with them, the fields associated with one bunch of particles can accelerate another bunch of particles. This simple idea is at the basis of the class of collective accelerators.

The plasma accelerator is a very promising tool to obtain larger accelerating fields than in conventional devices, capable to generate high-energy particles (and as well, neutral bunch of particles) with a relatively inexpensive apparatus, easily installed in any laboratory or university. Bostick [17] proposed that high energy ions could be accelerated by collective acceleration in order to simulate astrophysical plasmas. The understanding of the accelerating mechanism that propels and heat the solar wind is a decade-old mystery.

According to Alfvén & Carlqvist [18], it is crucial to introduce in the study of the cosmic plasma physics, the concepts of *electric double layers*, *critical velocity*, and *pinch effect*. One potential source of instability in a flow of charged particles is its constriction, the *Bennett pinch*. This phenomenon might explain the formation and abundance of filamentary structures at a cosmic scale. If we consider a fully ionized plasma in a cylindrical plasma column of radius r, in the presence of an axial electric field E_z that creates an axial current density j_z (source of an azimuthal magnetic field B_ϕ, the equation of conservation of momentum gives

$$\vec{\nabla} p = \vec{\nabla}\left(p_e + p_i\right) = \left[\vec{j} \times \vec{B}\right] \tag{13}$$

And then, using $\vec{\nabla} \times \vec{B} = \mu_0 \vec{j}$, together with the ideal gas law $p = NkT$, leads us to the Bennett relationship

$$2Nk\left(T_e + T_i\right) = \frac{\mu_0}{4\pi} I^2 . \tag{14}$$

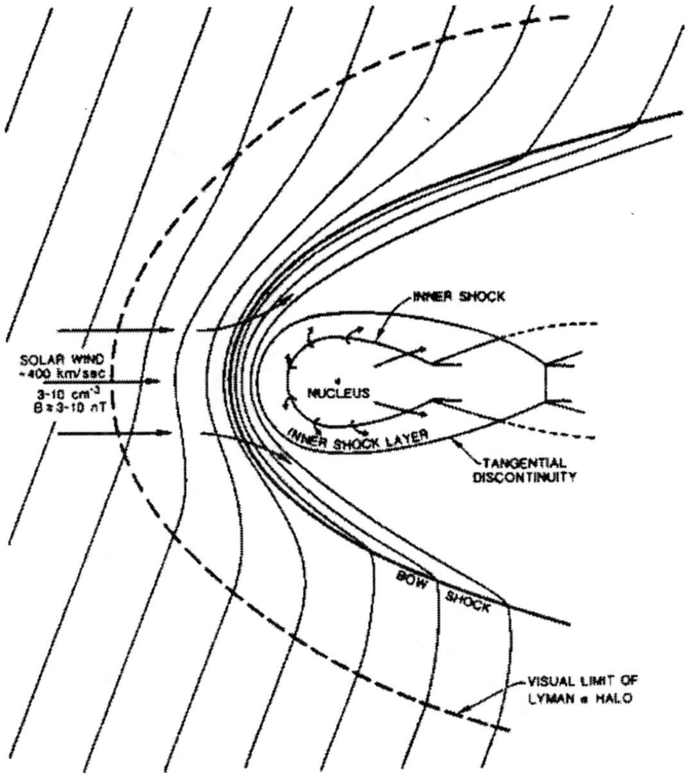

Figure 10. Typical structure resulting from the interaction of the solar wind with a comet. Image credit: Ref. [13].

Double layers are plasma structure, analogue to a cell wall that protects the plasma from the environment. They consist of two equal but oppositely charged space charge layers. In a glow discharge (produced between a cathode and anode at low pressure) it forms a cathode sheath and an anode sheath. Double layers are also important on the wall confining the plasma, by slowing down and decreasing the loss of electrons while speeding up the loss of positive ions. This means that the fluxes of ions and electrons, for example, in an unmagnetized plasma

$$\vec{\Gamma} = N\vec{u} = \mu N\vec{E} - D\vec{\nabla}N \, , \tag{15}$$

adjust themselves in order to the two species diffuse at the same rate. By setting it is obtained the ambipolar magnetic field and the ambipolar diffusion coefficient, $D_a = (\mu_i D_e + \mu_e D_i)/(\mu_i + \mu_e)$, where μ is the mobility, and the index i and e means ions and electrons.

Table 1. Typical magnitudes of relevant parameters in laboratory and cosmic plasmas. Adapted from Ref. [14]. l_c is a typical scale length of the plasma; ρ_m is the mass density; is the Lundquist number

	$l_c(m)$	$B(T)$	$\rho_m(kg/m^3)$	$\sigma(S/m)$	$V_A(m/s)$	L_u
Laboratory experiments						
Mercury	0.1	1	10^4	10^6	7.7	1
Sodium	0.1	1	10^3	10^7	30	37
Hydrogen	0.1	1	10^{-7}	5×10^4	2.8×10^6	2×10^4
Cosmic plasmas						
Earth's interior	10^6	10^{-3} (?)	10^4	8×10^5	10^{-2}	10^4
Sunspots	10^7	0.2	10^{-1}	4.4×10^4	20	1^{08}
Solar granulation	10^6	10^{-2}	10^{-4}	8×10^3	10^3	10^7
Magnetic variable stars	10^{10}	1	10^3(?)	8×10^5	30	3×10^{11}
Interstellar space	10^{20}	10^{-9}(?)	10-21 (?)	8×10^2 (?)	3×10^4	3×10^{21}
Interplanetary space	1011	10^{-8}	10^{-20}	8×10^4 (?)	10^5	10^{15}
Solar corona	10^9	10^{-4}(?)	10^{-15}(?)	8×10^5 (?)	3×10^6	3×10^{15}
Dark clouds	10^{11}	10^{-10}	10^{-17}	5×10^2	30	2×10^9

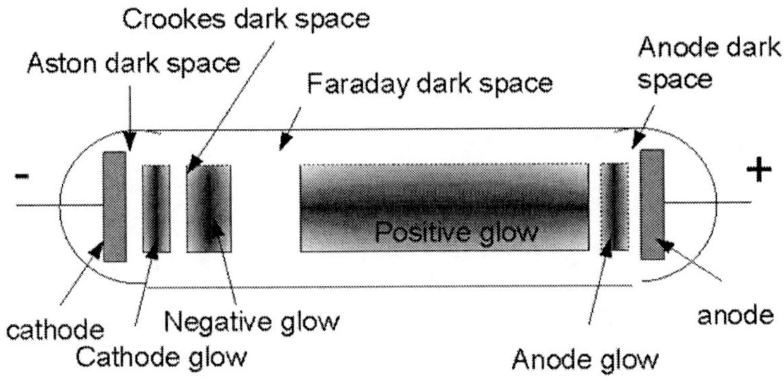

Figure 11. Different regions of a glow discharge.

There is an important relation that defines the electron temperature required to maintain the positive column (see Figure 11) in a steady-state,

$$\nu_i(\theta_e)/D_a(\theta_e) = (2.405/R)^2 . \tag{16}$$

Here, θ_e is the electron temperature and ν_i is the ionization excitation frequency. In a double layer (DL), the potential drop ϕ_0 along the layer obey to the relation, $|\Phi_0| \geq kT/e$, with T denoting the lowest temperature of the bordering region, and the other constant have the usual meaning. Inside the double layer the electric field is strong but does not exists outside it, and if the collision frequency is too high a double layer is not formed.

Space plasmas have typically electric field strengths of the order 1 V/m and S3-3 and Viking satellite data gives evidence of the existence of a huge number (hundreds or thousands) of aligned weak double layers whose total contribution can significantly increase the energy the accelerated particles. Figure 12 shows a sketch of the potential profile Φ; the current-carrying streaming electrons (n_-^s) and ions (n_+^s); the trapped electrons (n_-^t) and ions (n_+^t) at the downstream side; this group of four populations of particles is required to generate the net space-charge distribution ρ_c necessary to create the double layer potential.

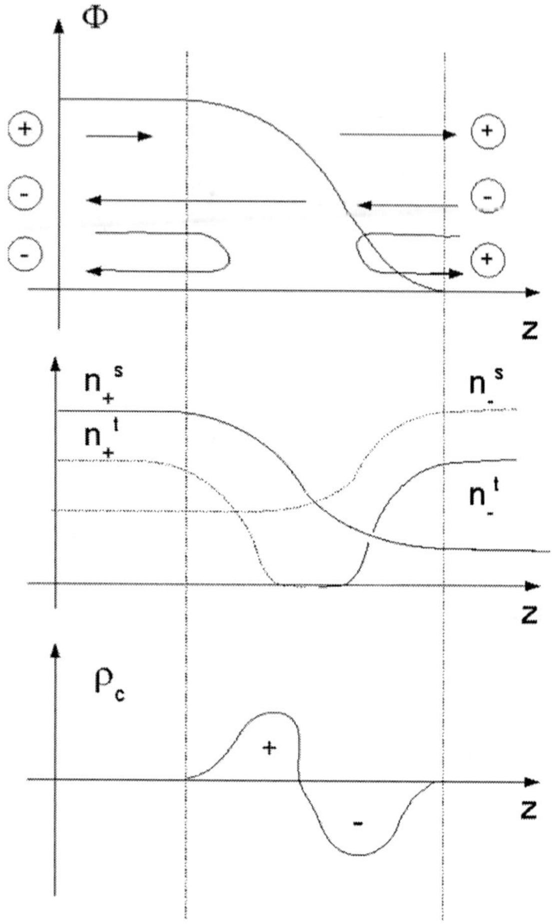

Figure 12. Sketch of the potential, the trapped (t) and streaming (s) ions (+) and electrons (-), the net space charge distributions through a double layer.

The streaming particles are accelerated by the potential drop inside the double layer, therefore creating the population of trapped particles on the downstream side. Particles crossing the DL are directly accelerated by the net potential Φ_0, the DL behaving as an electrical load dissipating energy at a rate $I\Phi_0$, where I is the total current across the DL, which by its turn is converted into kinetic energy of the accelerated particles (contrary to what happens in an ohmic resistance which would transform the electrostatic energy into heat). While in an electrical circuit what drives an electric current I is an electrical power source, in cosmic plasmas the driving mechanism is the variation of the energy stored in the magnetic field. A simple model can be considered in which an electric circuit contains a plasma filament with length l, radius of cross section a, current density j_0 and inductance of the order $L = \mu_0 l$. The magnetic energy stored in this circuit is $W_B = LI_0^2/2 = \mu_0 l I_0^2/2$, where the total filamentary current is $I_0 = \pi a^2 j_0$. The DL communicate to the particles the power $P_{DL} = I_0 \Phi_{DL}$ in a time scale of $\tau_B = W_B / P_{DL} = \mu_0 l I_0 / 2\Phi_{DL}$.

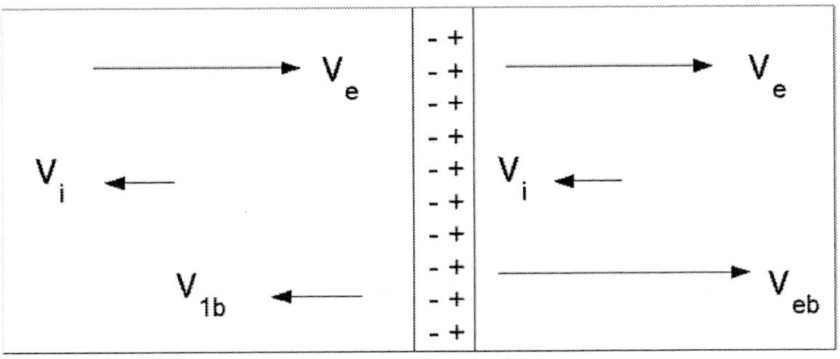

Figure 13. Velocity of electrons v_e and ions in a double layer, and resulting beam electron velocity v_{eb} and ion beam velocity v_{ib}.

Across the DL ions and electrons are accelerated, forming beams when leaving the layer, see Fig. 13, and exerting pressure upon the adjacent layers, eventually generating explosive DL.

In general terms, a Birkeland current is an electric current that follows along magnetic field lines. In magnetohydrodynamics equilibrium, the pressure gradient is negligible, being valid approximately the equation:

$$\vec{J} \times \vec{B} = 0 \tag{17}$$

This implies that the electric current must follow along the magnetic field lines. If we take into account the pressure gradient, such as

$$\nabla p = \left[\vec{J} \times \vec{B} \right] \tag{18}$$

and the Maxwell's equation $\vec{\nabla} \times \vec{B} = \mu_0 \vec{J}$, it is obtained

$$\vec{\nabla}\left(p+\frac{B^2}{2\mu_0}\right)=\frac{1}{\mu_0}\left(\vec{B}\vec{\nabla}\right)\vec{B} \tag{19}$$

In cylindrical coordinates (r,Φ,z), and assuming that the variables $p,\vec{B},\vec{J}$ depend only on the radial coordinate, we obtain from Eq.17

$$\frac{dp}{dr}=J_\phi B_z - J_z B_\phi \tag{20}$$

When the magnetostatic pressure is constant, $p=const.$, we have the condition

$$\frac{J_z}{J_\phi}=\frac{B_z}{B_\phi}=\tan\theta=K\,, \tag{21}$$

which leads to a field force-free condition of a spiral magnetic rope. In Eq.19, K and θ are functions of the radius r, while θ is the angle between the magnetic field $\vec{B}$ and the z-direction. Calculations show [19] that the distribution of the axial magnetic field B_z is confined inside a cylinder with a radius $b=B_0c/4\pi i_0$, while the current is constricted to a cylindrical region, forming a line current.

At larger distances it remains essentially the B_ϕ component (the B_z component diminishing radially), the overall structure is illustrated in Figure 14, from where it comes the concept of ´magnetic rope'. These structures are the backbone of the material filaments (observable mainly by means of the radiation they emit and absorb) that occur everywhere, in particular, in interstellar clouds and coronal filaments.

So, currents parallel to a magnetic field are prone to an instability that forces them to pinch into filaments and not flow homogeneously. At their pinch points an electric field appears, and small whirlpools are produced, adjacent filaments are attracted to these points where they cross, emitting here synchrotron radiation, and they combine into larger spirals. This is an important contribution of the plasma universe theory to cosmogony, providing strong evidence that plasmas play a fundamental role in the formation of the universe. But inhomogeneities have a universal role in plasma-dominated dynamics. According to Alfvén and Arrhenius [20], "Nature does not always have a *horror vacui* but sometimes *a horror homogeneitatis* resulting in an *amor vacui*".

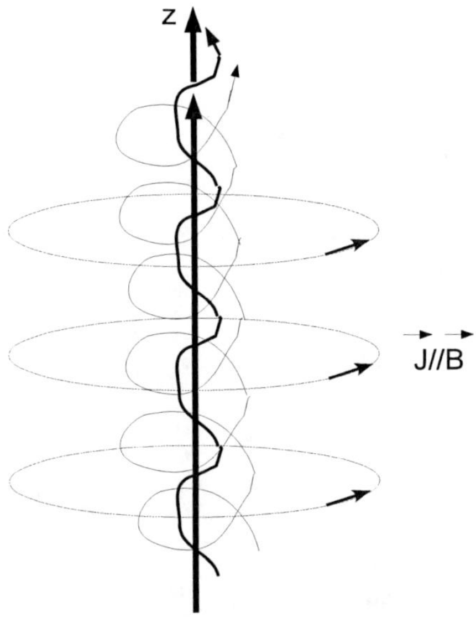

Figure14. A spiral trajectory in a force-free configuration with **J** and **B** parallel. Currents flow parallel to the magnetic field, having maximal magnitude at the center and decreasing of intensity radially.

The big bang gravitational model is at the heart of present cosmology, stating that the universe originates some fifteen billions to twenty billion years ago in a primordial explosion, ejecting matter and energy in all directions of space. Astronomers usually cite three strong points of the big bang cosmology: the expanding universe, as Edwin Hubble seemed to observe distant galaxies receding from Earth, in his Mount Wilson Observatory's 100-inch reflecting telescope; cosmic abundance of elements; the cosmic background radiation. Byproducts of this model are dark matter and black holes. But, with time, a growing number of unexplained phenomena are summing up: numerous cosmological objects with red shifts as intrinsic properties; the motion and shapes of galaxies cannot be explained only with the action of the gravitational force. Christian Birkeland, Irving Langmuir, Hannes Alfvén, Winston Bostick, and Anthony Peratt, are among the strongest supporters of the plasma universe theory.

The Faraday disk dynamo has a huge importance in our technological civilization, but this mechanism also occurs in astrophysics when conductive plasma moves in the cosmic magnetic field. They are known to power aurorae on Uranus, sunspots, binary stars, the Jupiter-Io system, and the Earth, for example.

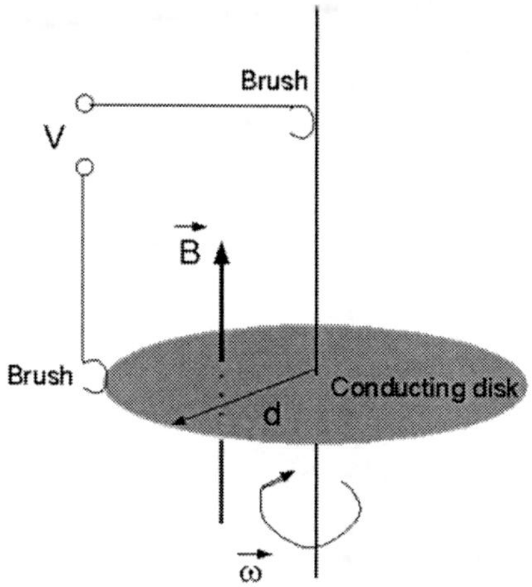

Figure 15. The Faraday disk dynamo. The conducting disk (or the cosmic plasma) with radius d turns with angular speed ω in the presence of a magnetic field B, appearing at the terminals a potential V. Mechanical energy can be converted into electric energy, and vice-versa.

Figure 15 illustrates the concept of the Faraday disk. A circular disk with radius d turns with angular speed in the presence of a magnetic field $\vec{B}$. Electrical contact with the axis and the disk boundary are made with brushes. An electron at a radial distance r from the center has a velocity ωr and it is actuated by the Lorentz force $e\omega rB$ directly radially outward. The electric field balancing the electron is directed radially inward with magnitude ωrB. The potential (electromotive force) between the center and the disk periphery is given by

$$V = \int_0^d E(r)dr = -\omega B \int_0^d r dr = -\frac{\omega B d^2}{2}. \quad (22)$$

When an electric field is present in the plasma, and a magnetic field acts perpendicularly, a radial inward convection of charged particles can be onset, under the influence of the $\vec{E} \times \vec{B}$ force, where both the ions and electrons drift with the velocity

$$\vec{v} = \frac{\vec{E} \times \vec{B}}{B^2}. \tag{23}$$

This mechanism is very effective in the accumulation of matter from plasma, the material acquiring a filamentary structure about the twisted magnetic flux tubes, in a kind of 'magnetic rope' pattern. In fact, a network of filamentary currents pervades our universe (Figure 1), they are subject to the pinch effect (an effect discovered in 1934 by Willard H. Bennett) and transfer energy and linear momentum over large distances, frequently producing electric double layers that, when arranged in linear configuration, may act like a kind of cosmic accelerators of highly energetic particles. Another mechanism with potential application in thermonuclear experiments is the concept of critical ionization velocity introduced by Alfvén. If the relative velocity between a magnetized plasma and a non-ionized gas overcomes a certain critical velocity $v_{crit} = \sqrt{2eV_{ion}/m_a}$, where V_{ion} is the ionization threshold and m_a is the atomic mass, then the interaction between both fluids are maximized, leading to a rapid ionization of the neutral gas [20]. We also may notice that the transition between fully ionized plasma and partially ionized plasma is often discontinuous. The Faraday homopolar motor shown in Figure 15 suggests that most of rotational motion of a plasma forms a generator of electromotive force between the center and periphery of the plasma vortices, and shows how amazingly nature find a way to obtain energy in our cosmos.

The standard interpretation of the sources of radiation coming up from galaxies, like the brightest synchrotron sources in the sky is Cygnus A, a double radio galaxy about 250,000 light-years apart, sustains that a black hole is the emitter of electron beams generating the synchrotron radiation. But this explanation is doubtful when compared with interpretation provided by the model of interacting plasma filaments. Also, computer simulations based on the plasma universe theory generate synchrotron radiation in a pattern amazingly similar to the ones observed in double radio galaxies. Dark matter theorists argue that the rotation of galaxies can be taken into account by their model. However, plasma cosmology can also explain, and replicate with computer models, the rotation of galaxies and the formation of the spiral arms in great detail [21]. These ideas form the body of a new cosmology based on the plasma universe model, from where it emerges the explanation of the origin of the sun, as formed from dusty interstellar cloud, and a huge number of astrophysical phenomena.

The hypothesis that the Sun may have an electric charge given by $Q_S = 5\times 10^{18}$ (C), was proposed by Prof. Bailey in 1960 for the explanation of the maximum energy found for a primary cosmic ray particle and other astronomical phenomena [22]. He considered that the four dimensional space-time universe was a hyper-surface in a five dimensional universe, and there exist streams of electrically charged particle from hyper dimensional universe into four dimensional space-time, where the laws of conservation of energy momentum and electric charge held true. He also thought that these electrical streams depend on local metric of space-time.

Thus we assume that his hypothesis is valid for the space-time continuum, the electric charge of the Sun can be estimated from the Gauss's law as

$$Q_S = \beta \int div\varepsilon_0 E \cdot dv = \beta\sqrt{4\pi\varepsilon_0 G}M_S \leq 1.7\times 10^{20} \text{ (C)}, \tag{24}$$

where is the mass of the Sun, and is an arbitrary constant satisfying $\beta \leq 1$.

This model considers that some of the electric charge of the Sun is induced by its own gravitational field. If there exist streams of charged particles from the hyper dimensional universe, as claimed by Prof. Bailey, the charge of the Sun generated by its own gravitational field is everlasting, as observed by T. T. Brown for the rock electricity, and the Sun may be powered by the current flow of electric charges attracted from plasma that stretches far out many times the radius of the Pluto by the massive electric charge of the Sun itself. In this view, the high temperature of the Sun's corona, which is considered to be heated by another mechanism than direct heat conduction from the photosphere, may be attributed to the electric currents induced by the charge of gravitational origin.

Boyko V. Ivanov, a researcher from the Institute for Nuclear Research and Nuclear Energy in Bulgaria, developed a theory that aims to describe the coupling between electromagnetism and gravitation, based on the well-known Weyl-Majumdar-Papapetrousolutions for the metric space-time [22-25].

In general relativity, EM fields alter the metric of space-time and induce a gravitational force through their energy-moment tensor given by

$$T_{\nu}^{\mu} = \frac{1}{4\pi}\left(F^{\mu\alpha} F_{\nu\alpha} - \frac{1}{4} \delta_{\nu}^{\mu} F^{\alpha\beta} F_{\alpha\beta} \right), \tag{25}$$

where is the electromagnetic tensor, and is the four potential. We have taken into account that $T_{\mu}^{\mu} = 0$, and assumed that the metric and the E-M fields do not depend on time. In this stationary case, let us simplify the problem by setting $A_{\mu} = (\overline{\phi},0,0,0)$, from where it may be concluded that there is just an electric field given by $E_{\mu} = F_{0\mu} = -\overline{\phi}_{\mu}$

Let us further assume that the space-time is static, then $f \equiv g_{00} = F(\phi)$ has the unique form $f = 1 + B\phi + \phi^2$, which was found by Weyl already in 1917 in the axially-symmetric case, which solutions are known as Weyl fields.

Then the equation for the gravitational field induced by static electric field given by

$$g = c^2 f^{-1} \left(\frac{B'}{2} \sqrt{\frac{\kappa\varepsilon}{8\pi}} \overline{\phi}_i + \frac{\kappa\varepsilon}{8\pi} \overline{\phi}\overline{\phi}_i \right) \tag{26}$$

where $f \equiv g_{00}$, is a constant, and $\kappa = 8\pi G / c^4$. From Eq.26 he derived the formula of gravitational force shown as

$$F_g = \sqrt{G\varepsilon} \frac{M}{d} \overline{\psi}_2 = \sqrt{G\varepsilon} \mu S \overline{\psi}_2 , \tag{27}$$

where is the mass of the dielectric, is its mass density, is dielectric constant, is the distance between the plates, is the potential of the second plate when $\psi_1 = 0$, and is an area of the plate. This effect has been discovered by Thomas Townsend Brown (1905-1985) already in 1923, together with Prof. P.A. Biefeld effect (now called the Biefeld-Brown effect).

Ivanov also pointed out that magnetic fields would produce the same effects as electric ones, from the definition of the magnetic field given by $H_i = -\frac{1}{2}\sqrt{-g}\varepsilon_{ikl}F^{kl}$.

Analogous to those for electric fields, he obtained the formula given by $g_i = -\sqrt{G\mu}H_i$, where denotes the magnetic constant. Thus one can increase the acceleration by creating magnetic fields in a ferromagnetic medium. But these electromagnetic-gravitational coupling effects are static, and thus it is considered that they can produce net force to propel the spaceship.

In this Chapter, we intended to introduce the background material with which we may build a new vision of the universe, well beyond the paradigmatic cosmological models based on the dominance of matter and gravitation. Much work has to be done in order to incorporate new data, in particular the wealth of new information provided by spacecraft observations, and, as well, a systematic deepening of our understanding of the working mechanism of our universe without *parti pris*.

CONCLUSION

The cosmic plasma theory is a different scientific approach to the workings of our universe that relies on a fundamentally different base than the orthodox cosmology, based on the dominance of gravitational forces, not duly recognizing the fundamental role of magnetic and electric fields on the formation of the galaxies and its cellular organization. It is an entirely different view of the nature and evolution of the universe that relies on data collected on the entire electromagnetic spectrum, starting from the observed fact that 99.999 percent of the universe is plasma, a universe dominated by electrical currents and electromagnetic forces that surpasses largely the gravitational force in magnitude.

These facts do not exclude the validity of the general theory of relativity and related disciplines, which are very useful to explain the behavior of matter in their advanced phase of existence, not in their origin. According to Tsiolkovsky's formula, a chemical-fuel rocket is not suitable for space flights and new solutions may be supplied by plasma physics.

REFERENCES

[1] Fälthammar, C.-G. Electrodynamics of Cosmical Plasmas-Some Basic Aspects of Cosmological Importance, *IEEE Transactions on Plasma Science*, 1990, 18, 11-17.

[2] Pinheiro, Mario J. *Plasma: the genesis of the word,* 2012, *PHILICA.COM Article number 363.*

[3] Alfvén, Hannes O. G. IEEE Trans. Plasma Sci.1990 18 (1) 5-10.

[4] Bohm, D. *Wholeness and the Implicate Order*; Routledge Classics: New York, NY, 1980.

[5] Berkovich, S. Obtaining inexhaustible clean energy by parametric resonance under nonlocal clocking, www.chronos.msu.ru/RREPORT/ berkovich_prime _energy.pdf (21.Sept.2010)

[6] Mandelstam, L.; Papaleksi, N. *Parametric excitation of electric oscillations* 1934 Technical Physics Journal Vol. IV (1)

[7] Mandelstam, L.; Papaleksi, N.; Andronov, A.; Chaikin, S.; Witt, A. 1935 *Exposé des recherches recentes sur les oscillations non-lineaires*, Technical Physics of the USSR, Leningrad; Vol. 2 (2-3), 81-134.

[8] Fossen, Thir I.; Nijmeijer, Henk *Parametric Resonances in Dynamical Systems*; Springer: New York, NY, 2012.

[9] Erturk, Alper; Inman, Daniel J. *Piezoelectric Energy Harvesting*; John Wiley & Sons, 2011.

[10] *Synergetics, a Workshop*, Proceedings of the International Workshop on Synergetics at Schloss Elmau, Bavaria, May 2-7, Edited by H. Haken, 1977.

[11] Lukash, V. N. Concealed Parameters of Dar Matter and Galaxy Formation, p. 331, in Morphological Cosmology, *Proc. Of the XIth Cracow Cosmological School Held in Cracow*, Poland, August 22-31, Eds. P. Flin and H. W. Duerbeck, 1988.

[12] Nordholt, J.E. et al. *Geophys. Res. Lett.* 2003 30(9) 1465.

[13] Russell, C. T. The Magnetosphere, pp. 73-100, in The Solar Wind and the Earth, edited by S.-I. Akasofu and Y. Kamide, Terrapub, Tokyo, 1987.

[14] Peratt, Anthony, L. Physics of the Plasma Universe; Springer-Verlag, NY, 1992.

[15] Alfvén, H.; Wernholm, P. *Arkiv Fysik* 1952 5, 175.

[16] Lawson, J. D. *Particle Accelerators* 1972 3 21.

[17] Bostick, W. H. What laboratory-produced plasma structures can contribute to the understanding of cosmic structures both large and small *IEEE Trans. Plasma Sci.* 1986 1 703.

[18] Alfvén, H.; Carlqvist, P., *Astrophys. Space Sci.* 1978, 55, 487-509.

[19] Alfvén H.; Fälthammar, C.-G. Cosmical Electrodynamics-Fundamental Principles; Oxford University Press: London, 1963.

[20] Alfvén, H.; Arrhenius, G. Report NASA-SP-345.

[21] Peratt, Anthony, L. The Sciences 1990 January/February.

[22] V.A.Bailey, Existence of Net Electrical Charges on Stars, *Nature* 1960, 186; 508-510.

[23] Ivanov, B.V., Strong gravitational force induced by static electromagnetic fields, arXiv: gr-qc/0407048 v1, July 13, 2004.

[24] Ivanov, B.V., On the gravitational field induced by static electromagnetic sources, arXiv: gr-qc/0502047 v1, February 10, 2005.

[25] Ivanov, B.V., Weyl Electrovacuum Solutions and Gauge Invariance, arXiv:gr-qc/0507082 v1, July 19, 2005.

Chapter 7

ELECTROGRAVITICS RESEARCH AND ADVANCED PROPULSION SYSTEMS

Thomas F. Valone[*]
Integrity Research Institute, Beltsville, MD, US

ABSTRACT

Research, based upon Thomas Townsend Brown's hypotheses, includes the idea that electrogravitics could be used as a means of propulsion for aircraft and spacecraft. Electrogravitic processes use a high voltage electric field to charge or, more properly, polarize an object with a specially-constructed shape. In this chapter, further research on electrogravitics and advanced propulsion systems that other scientists have produced are compared with Brown's inventions. The discovery by this author of the existence of an electrokinetic equation sheds light onto the basic parameters that affect and optimize the performance of electrogravitics and electrokinetics. The remarkable equation also provides an explanation for several heretofore unexplained propulsive experiments and predicts the direction of motion for them when the polarity of applied electrical charge is known.

1. INTRODUCTION

A curious fact revealed in T.T. Brown's first article [1] is the alignment of the "molecular gravitors". These massive dielectrics provided the most propulsive force when the "differently charged elements" were aligned (with the voltage source). This sounds like crystal plane alignment and perhaps explains the article "Gravity Nullified: Quartz Crystals Charged by High Frequency Currents Lose Their Weight" which appeared two years earlier in the same magazine in September of 1927. The editors had a change of heart however, in the following issue, and rescinded the article. Much of what we know about T.T. Brown is from his numerous patents (all of them are reprinted in the *Electrogravitics Systems* book [2]), although I was fortunate enough to correspond with him in 1981 when he was at the University of Florida. A sample of his detailed correspondence is contained in the book, *Ether-Technology: A Rational Approach to Gravity-Control* by Rho Sigma (1977) which is the only other introduction to Brown's work. A five-minute Brown-Bahnson Lab video online shows many of the experimental models that Brown developed toward 1960 [3] with Agnew Bahnson. The important fact from that book is that the DC power supply was raised substantially up to 250 kV, with a substantial force being displayed starting around 150 kV. Here we get an idea of the range of voltage necessary for successful electrogravitics that even recent military contractors mysteriously disregard. An example is R. L. Talley's report commissioned by the Air Force and concerned "with exploring the Biefeld-Brown effect which allegedly converts electrostatic energy directly into a propulsive force in a vacuum environment." It was entitled, "Twenty First Century Propulsion Concept" (#PL-TR-91-3009), but only tested Brown saucer designs in the range of 19 kV and predictably failed to produce any significant results.

Brown's saucer tests show a propulsive force with the positive voltage leading and the negative edge trailing. The high voltage electrically charged the air around the craft with a cloud of positive ions forming in front of the craft and a cloud of negative ions behind. This has been verified with tests recently performed by researcher Larry Davenport. These tests are reprinted in the book, *Electrogravitics Systems* and can also be seen in the commercial video, "Free Energy, The Race to Zero Point" for which I was the technical consultant.

Townsend Brown offered some interesting examples of advanced technology applications of electrokinetic designs in some of his patents. T.T. Brown's first patent, #1,974,483 issued in 1934 and was entitled "Electrostatic Motor." It is a fascinating free energy machine as well as a propulsion source. Claiming an efficiency of a "million to one", Brown causes the massive dielectrics to be the workhorse of the motor, exceeding, in his words, "the well-known pin wheel effect or reaction from a high voltage point discharge."

2. Further Electrogravitics Research

(1) Zinsser Effect versus the Biefeld-Brown Effect

There is another very similar invention which has comparable experiments that also involve electrogravity. It is the discovery of "gravitational anisotropy" by Rudolf G. Zinsser from Germany. I met with Zinsser twice in the early 1980's and corresponded with him subsequently regarding his invention. He presented his experimental results at the Gravity Field Conference in Hanover in 1980, and also at the First International Symposium of Non-Conventional Energy Technology in Toronto in 1981 [4]. For years afterwards, all of the scientists who knew of Zinsser's work regarded his invention as a unique phenomenon, not able to be classified with any other discovery. However, upon reading Brown's 1929 article on gravitation referred to above, I find striking similarities.

Zinsser's discovery is detailed in *The Zinsser Effect* book edited by this author [5]. To summarize his life's work, Zinsser discovered that if he connected his patented pulse generator to two conductive metal plates immersed in water, he could induce a sustained force that lasted even after the pulse generator was turned off. The pulses lasted for only a few nanoseconds each [6].

Zinsser called this input "a kinetobaric driving impulse." Furthermore, he points out in the Specifications and Enumerations section, reprinted in my book, that the high dielectric constant of water (about 80) is desirable and that a solid dielectric is possible. Dr. Peschka calculated that Zinsser's invention produced 6 Ns/Ws or 6 N/W [7].

This figure is *twenty times* the force per energy input of the Inertial Impulse Engine of Roy Thornson, which has been estimated to produce 0.32 N/W [8]. By comparison, it is important to realize that any production of force today is extremely inefficient, as seen by the fact that a DC-9 jet engine produces only 0.016 N/W or 3 lb/hp (fossil-fuel-powered land and air vehicles are even worse.)

Let's now compare the Zinsser Effect with the Biefeld-Brown Effect, looking at the details. Brown reports in his 1929 article that there are effects on plants and animals, as well as effects from the sun, moon and even slightly from some of the planetary positions. Zinsser also reports beneficial effects on plants and humans, including what he called "bacteriostasis and cytostasis."[9]

Brown also refers to the "endogravitic" and "exogravitic" times that were representative of the charging and discharging times.

Once the gravitator was charged, depending upon "its gravitic capacity" any further electrical input had no effect. This is the same phenomenon that Zinsser witnessed and both agree that the *pulsed voltage generation* was the main part of the electrogravitic effect [10].

Both Zinsser and Brown worked with dielectrics and capacitor plate transducers to produce the electrogravitic force. Both refer to a high dielectric constant material in between their capacitor plates as the preferred type to best insulate the charge. However, Zinsser never experimented with different dielectrics nor higher voltage to increase his force production. This was always a source of frustration for him but he wanted to keep working with water as his dielectric.

(2) Electrically Charged Torque Pendulum of Erwin Saxl

Brown particularly worked with a torque (torsion) pendulum arrangement to measure the force production. He also refers the planetary effects being most pronounced *when aligned with the gravitator* instead of perpendicular to it. He compares these results to Saxl and Allen, who worked with an electrically charged torque pendulum [11]. Dr. Erwin Saxl used high voltage in the range of +/- 5000 volts on his very massive torque pendulum [12]. The changes in period of oscillation measurements with solar or lunar eclipses showed great sensitivity to the shielding effects of gravity during an alignment of astronomical bodies, helping to corroborate Brown's observation in his 1929 article.

The pendulum Saxl used was over 100 kilograms in mass [13]. Most interesting were the "unexpected phenomena" which Saxl reported in his 1964 *Nature* article (see Reference [12]). The positively charge pendulum had the longest period of oscillation compared to the negatively charged or grounded pendulum.

Dirunal and seasonal variations were found in the effect of voltage on the pendulum, with the most pronounced occurring during a solar or lunar eclipse. In my opinion, this demonstrates the basic principles of electrogravitics: high voltage and mass together will cause unbalanced forces to occur. In this case, the electrogravitic interaction was measurable by oscillating the mass of a charged torque pendulum (producing current) whose period is normally proportional to its mass.

(3) Electrogravitics Woodward-Nordtvedt Effect

Referring to mass, it is sometimes not clear whether gravitational mass or inertial mass is being affected. The possibility of altering the equivalence principle (which equates the two), has been pursued diligently by Dr. James Woodward, whose patents can be reviewed in the Patent Section of this book. His prediction, based on Sciama's formulation of Mach's Principle in the framework of general relativity, is that "in the presence of *energy flow*, the inertial mass of an object may undergo sizable variations, changing as the 2nd time derivative of the energy."[15] Woodward, however, indicates that it is the "active gravitational mass" which is being affected but the equivalence principle causes both "passive" inertial and gravitational masses to fluctuate [16]. With barium titanate dielectric between disk capacitors, a 3 kV signal was applied in the experiments of Woodward and Cramer resulting in symmetrical mass fluctuations on the order of centigrams [17]. Cramer actually uses the phrase "Woodward effect" in his AIAA paper, though it is well-known that Nordtvedt was the first to predict noticeable mass shifts in accelerated objects [18].

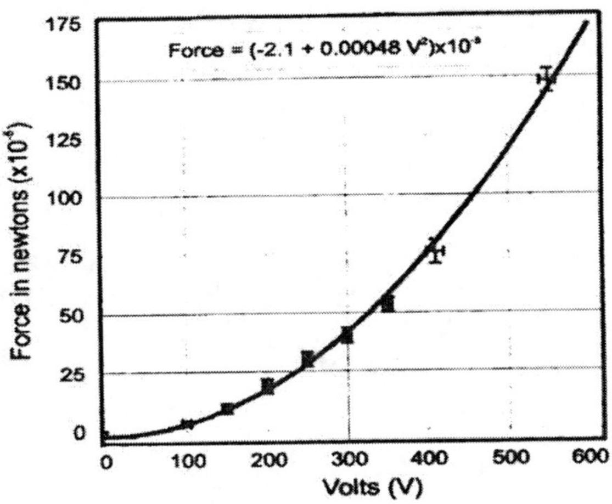

Figure 1. Force (10^{-5} N = dynes) output vs. capacitor voltage (V) input of a Woodward force transducer "flux capacitor" [14].

The interesting observation which can be made, in light of previous sections, is that Woodward's experimental apparatus *resembles a combination of Saxl's torsion pendulum and Brown's electrogravitic dielectric capacitors.* The differences arise in the precise timing of the pulsed power generation and with input voltage. Recently, 0.01 μF capacitors (Model KD 1653) are being used, in the 50 kHz range (lower than Zinsser's 100 kHz) with the voltage still below 3 kV. Significantly, the thrust or unidirectional force (UDF) is exponential, depending on the square of the applied voltage [19]. However, the micronewton level of force that is produced is actually the same order of magnitude which Zinsser produced, who reported his results in dynes (1 dyne = 10^{-5} Newtons). Zinsser had *activators* with masses between 200 g and 500 g and force production of "100 dynes to over one pound."[20] Recently, Woodward has been referring to his transducers as "flux capacitors" (like the movie, *Back to the Future*) [21].

(4) Jefimenko's Electrokinetic Equation Explains Electrogravitics

Known for his extensive work with atmospheric electricity, electrostatic motors and electrets, Dr. Oleg Jefimenko deserves significant praise for presenting a valuable theory of the *electrokinetic field*, as he calls it [22]. A West Virginia University professor and physics purist at heart, he describes this field as the dragging force that electrons exert on neighboring electric charges. He identifies the *electrokinetic field* by the vector $\mathbf{E}_k$, where

$$\mathbf{E}_k = -\frac{1}{4\pi\varepsilon_0 c^2}\int \frac{1}{r}\left[\frac{\partial \mathbf{J}}{\partial t}\right] dv' . \tag{1}$$

It is one of three terms for the electric field in terms of current and charge density. Equations like $\mathbf{F} = q\mathbf{E}$ also apply here for calculating force.

The significance of $\mathbf{E}_k$, as seen in Eq.(1), is that the electrokinetic field simply the third term of the classical equation for the electric field:

$$\mathbf{E} = \frac{1}{4\pi\varepsilon_0}\int \left\{\frac{\rho}{r^2} + \frac{1}{rc}\frac{\partial \rho}{\partial t}\right\} \mathbf{r} dv' + \mathbf{E}_k . \tag{2}$$

This three-term equation is a "causal" equation according to Jefimenko, because it links the electric field **E** back the electric charge and its motion (current) which induces it. This is the essence of electromagnetic induction, *as Maxwell intended*, which is measured by, not caused by, a changing magnetic field. The second electric field term, designated as the electrokinetic field, is directed along the current direction or parallel to it. It also exists only as long as the current is changing in time. Lenz' Law is also built into the minus sign. Parallel conductors will produce the strongest induced current.

By examining the vector potential **A** equation which depends upon the current density **J**, he finds that $\mathbf{E}_k$ can be expressed as the time derivative of **A**, which leads to

$$\mathbf{A} = -\int \mathbf{E}_k dt + const. \tag{3}$$

The significance of Eq.(3) is that the magnetic vector potential is seen to be created by the time integral which amounts to an *electrokinetic impulse* "produced by this current at that point when the current is switched on" according to Jefimenko [22], which resembles T.T. Brown's account of his experiments (see next section). Of course, a time-varying sinusoidal current will also qualify for production of an electrokinetic field and the vector potential. An important consequence of Eq. (1) is that *the faster the rates of change of current, the larger will be the electrokinetic force.* Therefore, high voltage pulsed inputs are favored. However, its significance is much more general. "This field can exist anywhere in space and can manifest itself as a pure force by its action on free electric charges." All that is required for a measurable force *from a single conductor* is that the change in current density (time derivative) happens very fast, to overcome the c^2 in the denominator.

The electrogravitics experiments of Brown and Zinsser involve a dielectric medium for greater efficacy and charge density. The electrokinetic force direction on the electric charges (electrons) of the dielectric, according to Eq. (1) which is a vector equation, is in the *opposite direction of the increasing positive current* (taking into account the minus sign). For parallel plate capacitors, Jefimenko explains that *the strongest induced field is produced between the plates* and so another equation evolves.

(5) Electrokinetic Force Predicts the Electrogravitic Direction of Motion

Can Jefimenko's electrokinetic force predict the correct *direction* of the electrogravitic force seen in the Zinsser, Brown, Woodward as well as the yet-to-be-discussed Campbell, Serrano, and Norton AFB craft demonstrations?

1) Starting with Zinsser's probe diagram (Figure 2) from Peschka's article, it is purposely put on its end for reasons that will become obvious. Compare it with an equivalent parallel plate capacitor (the plates are x distance apart) from Jefimenko's book [22].

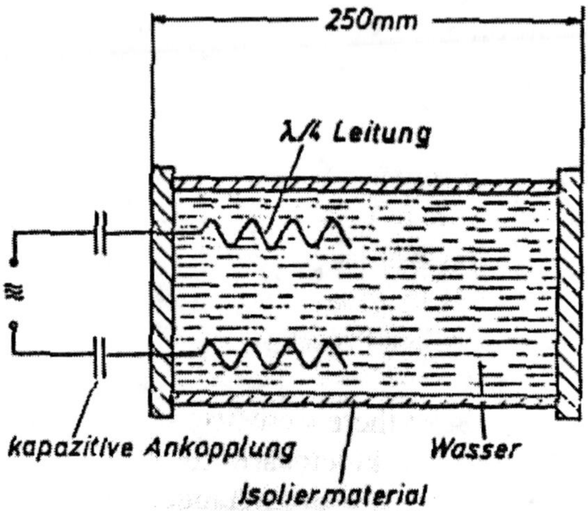

Figure 2. Sample capacitor probe used by Zinsser. Notice the quarter λ/4 wavelength electrodes which indicate a resonant circuit design.

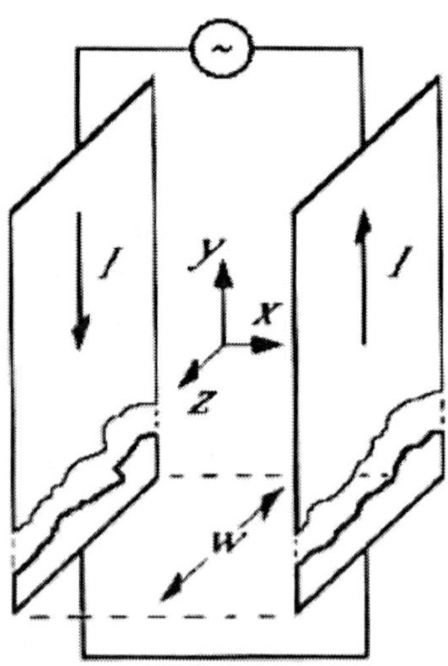

Figure 3. Calculation of Jefimenko's electrokinetic force in the space between two current-carrying plates with Eq. (4). X is the space between the plates. W is the width of the plates.

We note that the current is presumed to be the same in each plate but in opposite directions because it is alternating. Using Eq.(3), Jefimenko calculates the electrokinetic field, for the AC parallel plate capacitor with current going in opposite directions, as

$$E_k = -\mu_0 \frac{\partial I}{\partial t} \frac{x}{w} \mathbf{j}. \tag{4}$$

Of course, in vector calculus, **i**, **j**, and **k** are the unit vectors for the x, y, z axis directions seen in Figure 3, respectively. It is clearly seen that the y-axis points upward in Figure 3 and so with the minus sign of Eq.(3), the electrokinetic force for the AC parallel plate capacitor *will point downward.* Since Zinsser had his torsion balance on display in Toronto in 1981, I was privileged to verify the direction of the force that is created with his quarter-wave plates oriented as they are in Figure 2. The torsion balance is built so that the capacitor probe can only be deflected *downward* from the horizontal. *The electrokinetic force is in the same direction.*

2) Looking at Brown's electrogravitic force direction from the 1st figure in his 1929 article "How I Control Gravitation," we see that the positive lead is on the right side of the picture. Also, the arrow below *points to the right* with the caption, "Direction of movement of entire system toward positive." Examining the electrokinetic force of Eq. 1 in this article, we note that the increasing positive current comes in by convention in the positive lead and points to the left. Therefore, considering the minus sign, the direction of the electrokinetic force will be *to the right.* Checking with 4th figure of the 1929 Brown article, the same *confirmation of induced electrokinetic force direction.*[23] Thus, with Zinsser's and Brown's gravitators, *the electrokinetic theory provides a useful explanation and it also accurately predicts the resulting force direction when the applied current or voltage polarity is known.*

It is also very worthwhile noting that Brown also indicates in that article, "when the direct current with high voltage (75 – 300 kilovolts) is applied, the gravitator swings up the arc ... but it does not remain there. The pendulum then gradually returns to the vertical or starting positions, even while the potential is maintained...Less than five seconds is required for the test pendulum to reach the maximum amplitude of the swing, but from thirty to eighty seconds are required for it to return to zero."

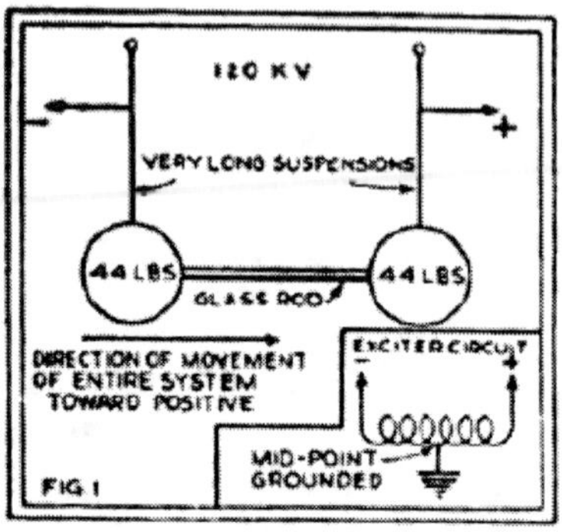

A SIMPLE TYPE OF GRAVITATOR IS SHOWN IN THE ABOVE ILLUSTRATION.

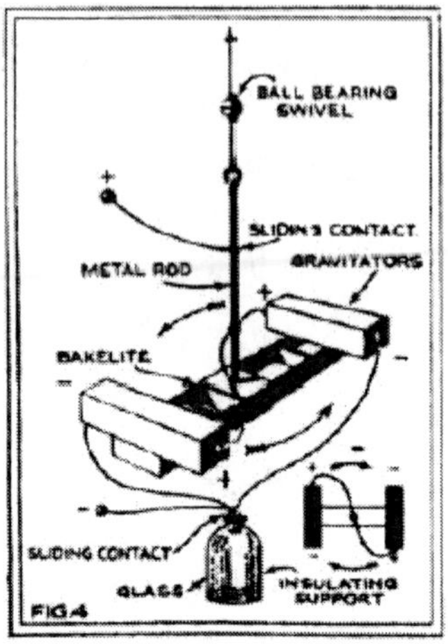

A GRAVITATOR ROTOR IS SIMPLY AN ASSEMBLY OF UNITS SO MADE THAT ROTATION RESULTS UNTIL THE IMPULSE IS EXHAUSTED.

Figure 4. Two drawings from T. Townsend Brown's famous article "How I Control Gravitation" which includes the polarity of the applied voltage and therefore applying the electrokinetic equation, a confirmation of the electrokinetic force direction can be obtained.

This candid account of the electrokinetic phenomena is *remarkably the same type of response that Zinsser recorded* with his experimental probes. Jefimenko's theory helps explain the rapid response, since the change of current happens in the beginning. However, the slow discharge in both experiments (which Zinsser called a "storage effect") needs more consideration. Considering the electrokinetic force of Eq. (4) and the +/- derivative, we know that the slow draining of a charged capacitor, most clearly seen in the 1st figure of Brown's 1929 article, will produce a decreasing current out of the + terminal (to the right) and in Eq.(2) or (4), this means *the derivative is negative.* Therefore, the slow draining of current will produce a weakening electrokinetic force but *in the same direction as before.* The force will thus sustain itself to the right during discharge.

3) It is very likely that the electrokinetic theory will also predict the direction of Woodward's UDF but instantaneous analysis needs to be made to compare current direction into the commercial disk capacitors and the electrokinetic force on the dielectric charges. In every electrogravitics or electrokinetics case, it can be argued, the "neighboring charges" to a capacitor plate will necessarily be those in the dielectric material, which are polarized. The bound electron-lattice interaction *will drag the lattice material with them,*

under the influence of the electrokinetic force. If the combination of physical electron acceleration (which also can be regarded as current flow) and the AC signal current flow can be resolved, it may be concluded that an instantaneous electrokinetic force, depending on dI/dt, contributes to the Woodward-Nordtvedt effect.

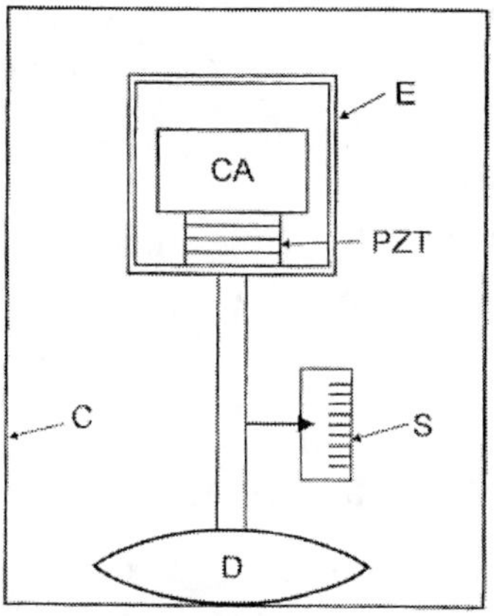

Figure 5. Woodward's #6,098,924 patented impulse engine, also called a flux capacitor. The PZT provides nanometer-sized movements that are timed to an AC signal input. A torsion balance has been used with a pair of force transducers in other designs.

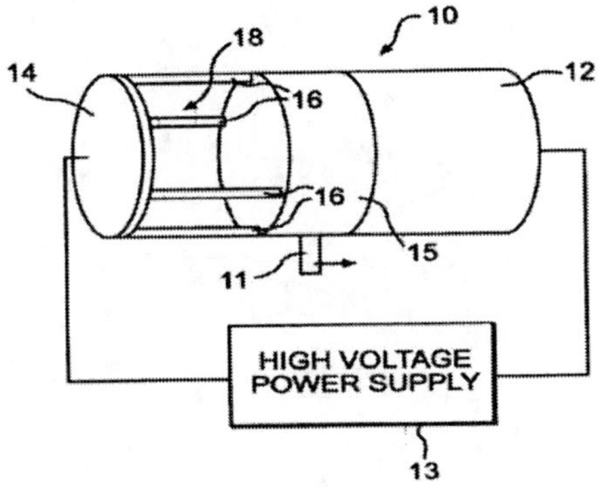

Figure 6. Capacitor module from Campbell's NASA patent #6,317,310 (above) which creates a thrust force as tested by Naudin. Disk 14 is copper; Struts 16 are dielectrics; Cylinder 15 is a dielectric; Cylinder 12 is an axial capacitor plate; Support post 11 is also dielectric. Compare with the Styrofoam and aluminum foil model below built by J. L. Naudin to test the Campbell NASA design, proving it produces thrust.

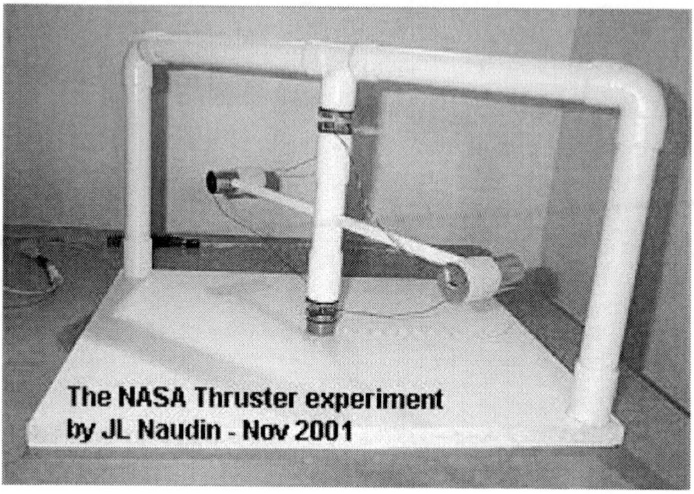

Figure 7. Styrofoam and aluminum foil model below built by J. L. Naudin to test the Campbell NASA design, proving it produces thrust.

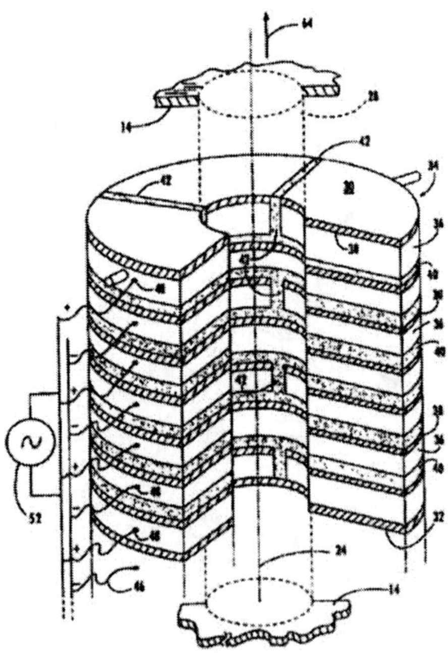

Figure 8. Capacitor propulsion device with alternating metal and dielectric layers from Serrano's PCT patent WO 00/58623 with upward thrust direction indicated and + and – polarity designated on the side, consistent with the electrokinetic force equation.

4) The Campbell and Serrano capacitor modules (Figures. 6 and 8), as well as the Electrogravitic Craft Demonstration unit (Figure 12), can all be explained with the electrokinetic force equation for the first time, in the same way that the Brown gravitator force was explained in paragraph (2) above. The current flows in one direction through the capacitor-dielectric and the force is produced in the opposite direction. The Norton AFB electrogravitic craft just has bigger plates with radial sections but the current flow still occurs at the center, *across the plates*. The Serrano patent diagram is also very similar in construction and operation.

(6) Electrokinetic Theory Observations

For parallel plate capacitor impulse probes, like Zinsser, Serrano, Campbell, the Norton AFB craft and both of Brown's models, the electrokinetic field of Eq.(4) provides a working model that seems to predict the *nature and direction of the force* during charging and discharging phases. More detailed information is needed for each example in order to actually calculate the theoretical electrokinetic force and compare it with experiment. We note that Eq. (4) also does not suffer the handicap of Eq. (1) since no c^2 term occurs in the denominator. Therefore, it can be concluded that AC fields operating on parallel plate capacitors should create significantly larger electrogravitic forces than other geometries with the same dI/dt. However, the current I is usually designated as $I_o\sin(\omega t)$ and its derivative is a sinusoid as well. Therefore, a detailed analysis is needed for each specific circuit and signal to determine the outcome. Eq.(4) also suggests a *possible enhancement* of the force if a permeable dielectric (magnetizable) is used. Then, the value for μ of the material would normally be substituted for μ_o [24].

A further observation of both Eq.(1) and Eq.(4) is that very fast changes in current, such as *a current surge or spark discharge* has to produce the most dynamic electrokinetic force, since dI/dt will be very large [25]. The declining current surge or the negatively sloped dI/dt however, should create an opposing force until the current reverses direction. *Creative waveshaping seems to be the answer* to this obvious dilemma. Fortunately, a few similar inventions use pulse power electric current generators to create propulsion.

The Taylor patent #5,197,279 "Electromagnetic Energy Propulsion Engine" uses huge currents to produce magnetic field repulsion. The Schlicher patent #5,142,861 "Nonlinear Electromagnetic Propulsion System and Method" predicts hundreds of pounds of thrust with tens of kiloamperes input.

The Schlicher antenna current input is a rectified current surge produced with an SCR-triggered DC power source (see Figure 9). The resulting waveform has a very steep leading edge but a *slowly declining trailing edge*, which should also be desirable for the electrokinetic force effect [26]. Furthermore, if this waveform is continued into the negative current direction below the horizontal axis, *all* of that region reinforces the electrokinetic force, with no opposite forces. Therefore, a complete sinusoidal wave, with Schlicher-style steep rise-times is recommended for a signal that contributes to a unidirectional force during 75% of its cycle.

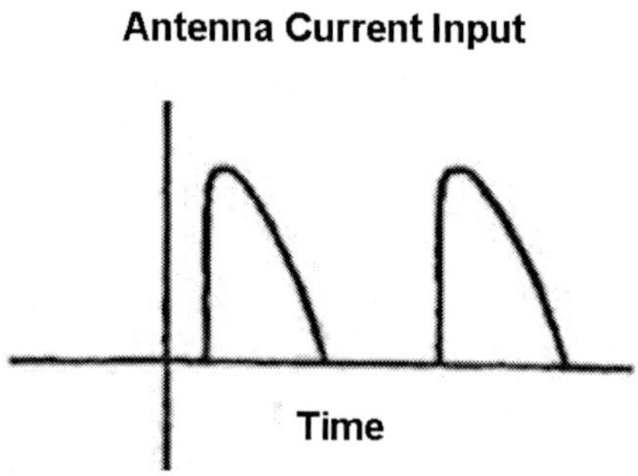

Figure 9. The ideal electrokinetic force current waveform is found in Schlicher propulsion patent #5,142,861. The same waveform is great for transporting calcium across cell membranes as well.

Another observation that should be mentioned is that this electrokinetic force theory does not include the mass contribution to the electrogravitic force which Saxl, Woodward, and Brown's 1929 gravitator emphasize. A contributor to the *Electrogravitics II* anthology and editor of this book, Takaaki Musha offers a derived equation for electrogravitics *that does include a mass term* but not a derivative term. His model is based on the charge displacement or "deformation" of the atom under the influence of a capacitor's 18 kV high voltage field and his experimental results are encouraging. He also includes a reference to Ning Li and her *gravitoelectric theory* [27].

A final concern, which may arise from the very nature of the electrokinetic force description, is the difficulty of conceptualizing or simply accepting the possibility of an *unbalanced force creation pushing against space*. This author has wrestled with this problem in other arenas for years. Three examples include (1) the homopolar generator which creates *back torque* that ironically, pushes against space to implement the Lorentz force to slow down the current-generating spinning disk [28]. Secondly (2), there is the intriguing *spatial angular momentum discovery* by Graham and Lahoz.[29] They have shown, reminiscent of Feynman's "disk paradox," that the vacuum is the seat of Newton's third law. A torsion balance is their chosen apparatus as well to demonstrate the pure reaction force with induction fields. Their reference to Einstein and Laub's papers cites the time derivative of the Poynting vector $\mathbf{S} = \mathbf{E} \times \mathbf{H}$ integrated over all space to preserve Newton's third law. Graham and Lahoz predict that *magnetic flywheels with electrets* will circulate energy to push against space. Lastly, for (3), the Taylor and Schlicher inventions push against space with an unbalanced force that is electromagnetic in origin.

3. Advanced Electrogravitic Systems

A fascinating article at the end of the 1950s surge in the gravity research is the report from 1961 in *Missles and Rockets* [30], which identifies a 389-page study released by the Office of Technical Services at the US Dept. of Commerce (may possibly be OTS #61-1187). The study however, sadly relates that disagreements among experts were becoming unyielding without more experimental proof.

Today, experimental proof seems to be in abundance. However, the prevailing trend by the government still fails to acknowledge the historical pioneering work of Biefeld and Brown, as well as any small inventor who is successful in this area. Take for example, Hector Serrano, who was interviewed in 1998 by NASA scientist Jonathan Campbell on video, about his electropropulsion invention. Within two years, Campbell started filing for a series of patents "on similar technology" but not referencing T.T. Brown nor Serrano in any of his US patents #6,317,310, #6,411,493, or #6,775,123 [31]. This type of behavior by a government representative is unethical and fuels the wide-spread public concern about government motives. Remarkably, it is like history repeating the same treatment that T.T. Brown received from the military upon demonstrating his working model to them [32]. A good seven-minute video summary of the Biefeld-Brown effect and the advanced technology of the B-2 is online at http://www.youtube.com/watch?v=8WsSXiEDLc8.

After all, aircraft are very massive and Brown's theory encourages the use of massive gravitators with high voltage, which we find in the B-2 bomber today. Paul LaViolette is credited for his discovery that the B-2 utilizes an auxiliary electrogravitics propulsion system with a flame jet generator, along with a million volt charge source for the leading edge of the wings (*Secrets of*

Antigravity Propulsion, Bear & Co., 2008). He documents how the B-2 advanced propulsion and stealth mode utilizes a high voltage polarity between the leading edge of the wings and the oppositely charged exhaust from a flame jet generator to create a separate system for forward thrust that can keep the plane aloft almost indefinitely.

Figure 10. The Northrop B-2 bomber which became available in 1994 and remains a technological wonder even today since it has flown missions to the Mid-East and back without refueling. Eye witnesses have reported the leading edge glows blue at night, apparently because it is using its high voltage electrogravitic propulsion.

In 2005, Deputy Director Bennett Hart of the National Reconnaissance Office gave a lecture at the National Space Society in Washington DC. When I approached him in the hallway on his way out, we discussed my conviction that *inertial shielding* is already available in classified projects and that it would help all efforts to launch spacecraft. He told me that it seems to be easier to direct contractors to develop technology *that he knows already exists*, mainly because declassification is very difficult. This is the main reason that we still use World War II technology on land and in space while the environment suffers irreparable harm. My sincere hope is that the validating science contained in *Electrogravitics II* will accelerate the civilian adaptation of this relatively simple propulsion technology.

The scientific articles in the *Electrogravitics II* anthology show the contrasting opinion that still exists in the assessment of electrogravitics. As *inertial shielding* also is reinvented by civilian scientists, I predict that electrogravitics will become more and more useful and a primary means of transportation in outer space. The reason behind my prediction is that any force moving a mass utilizes Newton's Second Law, F = ma, which can be very powerful when the inertial mass m is reduced by electrogravitic shielding. Once again, such a technological development already exists, as exhibited in the night photos of right-angle turns and fast acceleration of covert, black project triangular and circular craft which are operated by classified projects in America.

Figure 11. Advanced electrogravitic prototype of Godin and Roschin who obtained a US Patent #6,822,361for an orbiting homopolar generator based on the work of John R. R. Searl.

No one has publicly admitted to successfully replicating the legendary energy and antigravity device called the "Searl Effect Generator" (SEG) and "Inverse Gravity Vehicle" (IGV), developed in 1946 by John R. R. Searl. However, there has been a curious increase in worldwide sightings of circular flying objects since 1950, so rumors have it that many of these sightings are from terrestrial advanced aircraft. Whatever the explanation, the most likely propulsion mechanism for these craft is the application of rotating magnetic fields to gravity control possibly following the Searl design, which also has many references in the physics and patent literature. However, no commercial development has succeeded so far for various reasons.

In June, 2000, Sergei M. Godin and Vladimir V. Roschin, with the Russian Academy of Science, in Moscow, notified the U. S. Department of Energy (DOE) that some expert opinions were needed to understand the experimental phenomena from a magnetic energy converter (MEC) experiment that was based on the SEG and IGV. Their paper was published soon after in *Technical Physics Letters* [34] and also circulated by the DOE to a select number of recipients, including myself, in the Advanced EM Group for scientific feedback, to which I responded in the affirmative.

The experimental device fashioned after a Searl disk, required only a starter motor according to their published account, (1) became self-sustaining at 300 RPM, (2) successfully produced 35% loss of weight, (3) required 5-7 kilowatt energy drain to prevent a runaway condition, (4) caused a temperature decrease of 6-8 degrees Celsius in the local vicinity, (5) created anomalous magnetic zones up to 15 meters around the device, and (6) displayed unexplained luminous horizontal bars surrounding the device when charged to 20 kV of high voltage.

However, upon receiving a substantial investment to replicate the original device they tested, the Russian team still has not been able to obtain the same results as with the earlier device, even after more than ten years of work (see Figure 11). Many other groups around the world still hold out hope for the unusual device that Searl reported can generate electricity (SEG) and also be configured for propulsion (IGV). A self-published book by John Thomas [35] is a good illustrated introduction to the story of John R. R. Searl. It is worth mentioning the existence of twelve volumes of dictated works by Searl called "*The Law of the Squares*" edited by John Thomas which overwhelms the investigator with technical details and mathematical relationships that are purported to be critical for the IGV performance. The IGV is electrogravitic by nature since it is supposed to generated very high voltage when self-sustaining which can be controlled by unbalancing the charge on the circular craft edges to direct the motion toward the positive side, in the same way that the Biefeld-Brown effect works.

In other related, advanced research, Bernard Haisch, Harold E. Puthoff, and several other physicists [36] have shown intriguing connections between electromagnetics, notably the electromagnetic zero-point field, and inertia, and have speculated about possible further connections with gravity. Physicist Ning Li and engineer Eugene Podkletnov have, respectively, shown theoretically, and reported observing experimentally, anomalous gravitic attenuation effects above a superconducting disk spinning in a strong magnetic field such as is produced in a Meissner effect demonstration apparatus [37].

According to Li, rotating ions create a gravitomagnetic field perpendicular to their spin axis [38, 39]. If a large number of ions could be aligned, (in a Bose–Einstein condensate) the resulting effect would be a very strong gravitomagnetic field producing a strong repulsive force. The alignment may be possible by trapping superconductor ions in a lattice structure in a high-temperature superconducting disc. Giovanni Modanese has conducted further experiments on the phenomena seen by Podkletnov, and has reported some additional much stronger, but transient, anomalous gravitational effects [40, 41]. An example of an advanced propulsion system due to electrokinetics is due to Mark McCandlish, who offers us the conclusive perspective of the covert, flat-bottomed saucer hovercraft ("ARV") seen by dozens of invited eye-witnesses and a Congressman at Norton Air Force Base on November 12, 1988 (a complete description by Mark McCandlish is included in the last chapter of *Electrogravitics II* [33]).

In analyzing the Electrogravitic Craft Demonstration unit (Norton AFB 1988) diagrammed at the end of this chapter, I have compared it to Campbell's and Serrano's patented design. A lot can be learned from studying the intricacies of this advanced design. The ARV (Figure 12) is called an "Alien Reproduction Vehicle" or "Advanced Research Vehicle" including the use of a distributor cap style of pulse discharge in the center and multiple symmetric, radial capacitor plates with dielectrics in between.

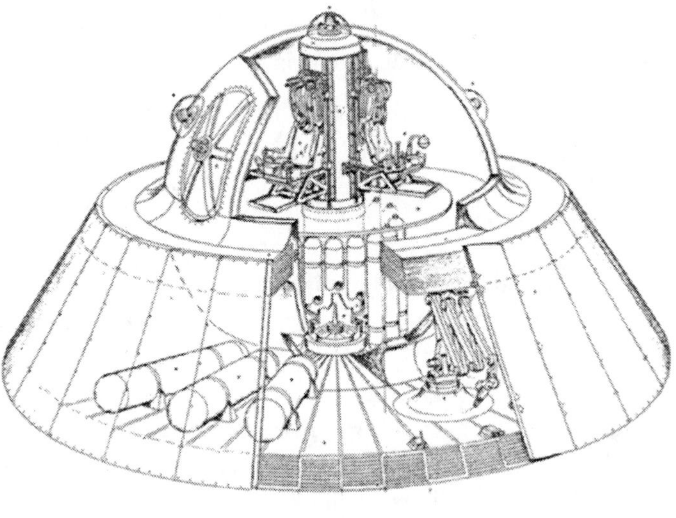

Figure 12. The amazing Advanced Research Vehicle (ARV) diagrammed by aerospace illustrator, Mark McCandlish, which is powered by pulsed electrokinetics as the central distributor cap rotates and fires three equidistant capacitor banks at a time.

Mark indicated that three capacitor banks equally spaced apart were fired at any given time to create a hovering effect that equaled the weight of the vehicle. He even called the Congressman's office the next day to thank the staffer for the demonstration and thereby also verify the awareness of the office that the advanced propulsion demonstration event took place.

CONCLUSION

With all of the advanced technology discussed in this chapter related to electrogravitics and electrokinetics, it is expected that civilian research will eventually catch up to the black project back-engineered propulsion systems that already exist. With climate change becoming more and more of a threat, the hope of utilizing electrogravitics exhaust-free propulsion is a dream come true. We can look forward to the day when a saucer craft is parked in every suburban garage, instead of the old-fashioned, 1930s-style, land-based fossil fuel burner on wheels we are all too familiar with. Then, instead of burning oil and causing tons of pollution to go into the atmosphere, we will be using an onboard quantum vacuum energy generator-battery system to generate high voltage with very little current, and along with the built-in inertial shield, we will have the freedom of leaving the atmosphere for a jaunt to the asteroid belt or to the moon for an afternoon drive in outer space. The same advanced propulsion that we envision within this book will also enable the mining of asteroids which have been estimated to contain trillions of dollars' worth of precious metals.

REFERENCES

[1] Brown, T.T., "How I Control Gravity", *Science and Invention*, 1929.
[2] Valone, Thomas, Electrogravitics Systems, Integrity Research Institute, Washington. D.C.2004.
[3] http://www.youtube.com/watch?v=Rp4hygoD3RU.
[4] Zinsser, R. G., Mechanical Energy from Anisotropic Gravitational Fields, First Int'l Symp. on Non-Conventional Energy Tech. (FISONCET), Toronto, 1981. Proceedings available from PACE, 100 Bronson Ave #1001, Ottawa, Ontario K1R 6G8.

[5] Valone, Thomas *The Zinsser Effect: Cumulative Electrogravity Invention of Rudolf G. Zinsser*, Integrity Research Institute.org , 2005, 130 pages, IRI #701.

[6] Cravens, D.L., Electric Propulsion/Antigravity, *Electric Spacecraft Journal*, Issue 13, 1994, p. 30.

[7] Peschka, W., Kinetobaric Effect as Possible Basis for a New Propulsion Principle, *Raumfahrt-Forschung*, Feb, 1974. Translated version appears in *Infinite Energy*, Issue 22, 1998, p. 52 and *The Zinsser Effect*.

[8] Valone, Thomas, "Inertial Propulsion: Concept and Experiment, Part 1" Proc. of Inter. Energy Conver. Eng. Conf., 1993, See IRI Report #608.

[9] Pulsed Electromagnetic Field Health Effects, IRI Report #418 and *Bioelectromagnetic Healing* book #414 by this author, which explains the beneficial therapy which PEMFs produce on biological cells.

[10] Mark McCandlish's Testimony shows that the Air Force took note in that the electrogravitic demonstration craft shown at Norton AFB in 1988 had a rotating distributor for electrically pulsing sections of multiply-layered dielectric and metal plate pie-shaped sections with high voltage discharges. (See p. 131, *Electrogravitics II, Validating Reports on a New Propulsion Methodology*, by Thomas Valone, Integrity Research Institute publishers, 2008)

[11] Saxl patent #3,357,253, Device and Method for Measuring Gravitational and Other Forces, which uses +/- 5000 volts.

[12] Saxl, E.J., "An Electrically Charged Torque Pendulum" *Nature*, July 11, 1964, p. 136.

[13] Saxl & Allen, "Observations with a Massive Electrified Torsion Pendulum: Gravity Measurements During Eclipse," IRI Report #702.(Note: 2.2 lb = 1 kg)

[14] Woodward and Mahood, Mach's Principle, Mass Fluctuations, and Rapid Spacetime Transport, California State University Fullerton, Fullerton CA 92634.

[15] Cramer et al., Tests of Mach's Principle with a Mechanical Oscillator, AIAA-2001-3908 email: cramer@phys.washington.edu.

[16] Woodward, James F., A New Experimental Approach to Mach's Principle and Relativistic Gravitation, *Found. of Phys. Letters*, V. 3, No. 5, 1990, p. 497.

[17] Compare Figure 1 graph to Brown's ONR graph on P.117 of *Electrogravitics Systems, Reports on a New Propulsion Methodology*, (Volume I) by Thomas Valone, Integrity Research Institute publishers, 1994.

[18] Nordtvedt, K. *Inter. Journal of Theoretical Physics*, V. 27, 1988, p. 1395.
[19] Mahood, Thomas "Propellantless Propulsion: Recent Experimental Results Exploiting Transient Mass Modification" Proc. of STAIF, 1999, CP458, p. 1014 (Also see Mahood Master's Thesis www.serve.com/mahood/thesis.pdf.
[20] Zinsser, FISONCET, Toronto, 1981, p. 298.
[21] Woodward, James, Flux Capacitors and the Origin of Inertia, *Foundations of Physics*, V. 34, 2004, p. 1475. Also see, Tweaking Flux Capacitors, *Proc. of STAIF*, 2005.
[22] Jefimenko, Oleg, *Causality, Electromagnetic Induction and Gravitation*, Electret Scientific Co., POB 4132, Star City, WV 26504.
[23] Brown's second patent #2,949,550 (two electrokinetic saucers on a maypole) has movement toward the positive charge, so the same electrokinetic theory explained above works for both.
[24] Einstein and Laub, *Annalen der Physik*, V. 26, 1908, p.533 and p. 541 – two articles on the subject of a moving capacitor with a "dielectric body of considerable permeability." Specific equations are derived predicting the resulting EM fields. Translated articles are reprinted in *The Homopolar Handbook* (Integrity Research Institute publishers, 2001) by Thomas Valone (p. 122-136). Also see Clark's dielectric homopolar generator patent #6,051,905.
[25] Commentary to Eq. (2) states an electrokinetic impulse is produced when the "current is switched on," which implies a very steep leading edge of the current slope.
[26] Taylor and Schlicher patents in the Patent Section.
[27] Ning Li was the Chair of the 2003 Gravitational Wave Conference. The CD Proceedings of the papers is available from Integrity Research Institute.
[28] Valone, Thomas, *The Homopolar Handbook: A Definitive Guide to Faraday Disk and N-Machine Technologies*, Integrity Research Institute, Third Edition, 2001.
[29] Graham and Lahoz, Observation of Static Electromagnetic Angular Momentum in vacuo, *Nature*, V. 285, May 15, 1980, p. 129.
[30] Beller, William, Soviet Efforts are Closely Watched, *Missiles and Rockets*, Sept. 11, 1961, p. 27.
[31] Young, Kelly, Inventor: NASA stole patent idea, *Florida Today*, Sept. 29, 2002 (entire article is posted on the *Florida Today* website).

[32] The Townsend Brown Electro-Gravity Device, File 24-185, A Comprehensive Analysis by the Office of Naval Research, Sept. 15, 1952, - IRI report #612.

[33] Valone, Thomas, Electro Gravitics II, Integrity Research Institute, Washington, DC, 2000.

[34] Sergei M. Godin, S.M., Roschin,*V. V., Technical Physics Letters* , Vol. 26, No. 24, 2000, p. 70.

[35] Thomas, J., *Antigravity: The Dream Made Reality*, 1993.

[36] Haish, B., Rueda, A. and Puthoff, H.E., Physics of the Zero-Point Field: Implications for Inertia, Gravitation and Mass, Spec. in Sci. and Technology 20, 1997, p.20.

[37] Torr D.G., Li N., Graviotoelectric-Electric Coupling via Superconductivity, Foundations of Physics, Vol.6, No.4; 1993, 371-383.

[38] Li N., Torr D.G., Gravitational effects on the magnetic attenuation of superconductors, Physical Review B, Vol.46, No.9 ; 1992, 5489-5495.

[39] Li N., Noever D.,.Robertson T., Koczor R., Brantley W., Static test for a gravitational force coupled to type II YBCO superconductor, Physica C, 281; 1997, 260-267.

[40] Podkletnov, E., Modanese, G., Impulse gravity generator based on charged YBa2Cu3O7-y superconductor with composite crystal structure, LANL Physics Preprint Server, arXiv: physics/0108005 (2001, August.3).

[41] Podkletnov, E., Modanese, G., Investigation of high voltage discharges in low pressure gases through large ceramic superconducting electrodes, Journal of Low Temperature Physics, 132(3-4), 2003, p239. Also, LANL Preprint Server, arXiv:physics/0209051.

INDEX

E

F

G

H

I

J

K

L

M

N

O

P

Q

R

S

Z